T0255977

Advances in Intelligent Systems and Computing

Volume 883

Series editor

Janusz Kacprzyk, Systems Research Institute, Polish Academy of Sciences, Warsaw, Poland
e-mail: kacprzyk@ibspan.waw.pl

The series "Advances in Intelligent Systems and Computing" contains publications on theory, applications, and design methods of Intelligent Systems and Intelligent Computing. Virtually all disciplines such as engineering, natural sciences, computer and information science, ICT, economics, business, e-commerce, environment, healthcare, life science are covered. The list of topics spans all the areas of modern intelligent systems and computing such as: computational intelligence, soft computing including neural networks, fuzzy systems, evolutionary computing and the fusion of these paradigms, social intelligence, ambient intelligence, computational neuroscience, artificial life, virtual worlds and society, cognitive science and systems, Perception and Vision, DNA and immune based systems, self-organizing and adaptive systems, e-Learning and teaching, human-centered and human-centric computing, recommender systems, intelligent control, robotics and mechatronics including human-machine teaming, knowledge-based paradigms, learning paradigms, machine ethics, intelligent data analysis, knowledge management, intelligent agents, intelligent decision making and support, intelligent network security, trust management, interactive entertainment, Web intelligence and multimedia.

The publications within "Advances in Intelligent Systems and Computing" are primarily proceedings of important conferences, symposia and congresses. They cover significant recent developments in the field, both of a foundational and applicable character. An important characteristic feature of the series is the short publication time and world-wide distribution. This permits a rapid and broad dissemination of research results.

More information about this series at http://www.springer.com/series/11156

Rituparna Chaki · Agostino Cortesi ·
Khalid Saeed · Nabendu Chaki
Editors

Advanced Computing and Systems for Security

Volume Eight

 Springer

Editors
Rituparna Chaki
A.K. Choudhury School of Information
Technology
University of Calcutta
Kolkata, West Bengal, India

Khalid Saeed
Faculty of Computer Science
Bialystok University of Technology
Bialystok, Poland

Agostino Cortesi
Dipartimento di Scienze Ambientali,
Informatica e Statistica
Università Ca' Foscari
Mestre, Venice, Venezia, Italy

Nabendu Chaki
Department of Computer Science
and Engineering
University of Calcutta
Kolkata, West Bengal, India

ISSN 2194-5357   ISSN 2194-5365   (electronic)
Advances in Intelligent Systems and Computing
ISBN 978-981-13-3701-7   ISBN 978-981-13-3702-4   (eBook)
https://doi.org/10.1007/978-981-13-3702-4

Library of Congress Control Number: 2018964250

This Springer imprint is published by the registered company Springer Nature Singapore Pte Ltd.
The registered company address is: 152 Beach Road, #21-01/04 Gateway East, Singapore 189721,
Singapore

Preface

The Fifth International Doctoral Symposium on Applied Computation and Security Systems (ACSS 2018) was held at Kolkata, India, during February 9–11, 2018. The University of Calcutta in collaboration with Ca' Foscari University of Venice, Italy, and Bialystok University of Technology, Poland, organized the symposium.

The fifth year of the symposium was marked with a significant shift of research interests toward usage of machine learning based on signal processing and image analytics. It is truly interesting to find the increasing popularity of ACSS among researchers, this being one of its own kinds of symposium for doctoral students to showcase their ongoing research works on a global platform. Program Committee Members are each renowned researchers in his/her field of interest, and we thank them for taking immense care in finding out the pros and cons of each of the submissions for ACSS 2018. As in the previous years, Session Chairs for each session had a prior go-through of each paper to be presented during the respective sessions. This often makes it more interesting as we found deep involvement of Session Chairs in mentoring the young scholars during their presentations. With concrete suggestion on how to improve the presented works, the participants utilized the 6-week post-symposium time. The final version of the papers thus goes through the second level of modification as per Session Chair's comments.

These post-symposium book volumes contain the revised and improved version of the manuscripts of works presented during the symposium. The evolution of ACSS is an ongoing process. We had included deep learning in the scope of research interests in 2018. In 2018, considering the global interest, we plan to include pricing and network economics within the scope and announced the same in CFP for the next year. We have invited researchers working in the domains of algorithms, signal processing and analytics, security, image processing, and IoT to submit their ongoing research works.

The indexing initiatives from Springer have drawn a large number of high-quality submissions from scholars in India and abroad. We are indebted to all Program Committee Members, who, despite their immensely busy schedules, have given an intense reading of each allotted contribution. Each reviewer has given their constructive suggestions for the submitted works. ACSS continues with the tradition of the double-blind review process by PC Members and by the external reviewers. The reviewers mainly considered the technical aspect and novelty of each paper, besides the validation of each work. This being a doctoral symposium, clarity of presentation was also given importance. The entire process of paper submission, review, and acceptance was done electronically. Due to the sincere efforts of Program Committee and Organizing Committee Members, the symposium resulted in a suite of strong technical paper presentations followed by effective discussions and suggestions for improvement for each researcher.

The technical program committee for the symposium selected only 24 papers for publication out of 64 submissions. We would like to take this opportunity to thank all Members of the program committee and the external reviewers for their excellent and time-bound review works. We thank all the sponsors who have come forward toward the organization of this symposium. These include Tata Consultancy Services (TCS), Springer Nature, ACM India, and M/s Fujitsu, Inc., India. We appreciate the initiative and support from Mr. Aninda Bose and his colleagues in Springer Nature for their strong support toward publishing this post-symposium book in the series "Advances in Intelligent Systems and Computing." Last but not least, we thank all the authors without whom the symposium would not have reached up to this standard.

On behalf of the editorial team of ACSS 2018, we sincerely hope that ACSS 2018 and the works discussed in the symposium will be beneficial to all its readers and motivate them toward even better works.

Kolkata, India   Rituparna Chaki
Bialystok, Poland   Khalid Saeed
Venice, Italy   Agostino Cortesi
Kolkata, India   Nabendu Chaki

Contents

Part IV   Image Processing

About the Editors

Rituparna Chaki is Professor of Information Technology in the University of Calcutta, India. She received her Ph.D. degree from Jadavpur University in India in System Executive in the Ministry of Steel, Government of India for nine years, before joining academia in 2005 as a Reader of Computer Science & Engineering at the West Bengal University of Technology, India. She has been with the University of Calcutta since 2013. Her areas of research include optical networks, sensor networks, mobile ad hoc networks, Internet of Things, and data mining. She has nearly 100 publications to her credit. Dr. Chaki has also served on the program committees of various international conferences, and has been a regular Visiting Professor at the AGH University of Science & Technology, Poland for last few years. Dr. Chaki has co-authored a couple of books published by CRC Press, USA.

Agostino Cortesi, Ph.D., is a Full Professor of Computer Science at Ca' Foscari University, Venice, Italy. He served as Dean of Computer Science Studies, as Department Chair, and as Vice-Rector for quality assessment and institutional affairs. His main research interests concern programming languages theory, software engineering, and static analysis techniques, with a focus on security applications. He has published more than 110 papers in high-level international journals and international conference proceedings. His h-index is 16 according to Scopus, and 24 according to Google Scholar. Agostino served several times as a member (or chair) of program committees of international conferences (e.g., SAS, VMCAI, CSF, CISIM, ACM SAC) and he is on the editorial boards of the journals "Computer Languages, Systems and Structures" and "Journal of Universal Computer Science". Currently, he holds the chairs of "Software Engineering", "Program Analysis and Verification", "Computer Networks and Information Systems" and "Data Programming".

Khalid Saeed is a Full Professor in the Faculty of Computer Science, Bialystok University of Technology, Bialystok, Poland. He received his B.Sc. degree in Electrical and Electronics Engineering from Baghdad University in 1976, and

M.Sc. and Ph.D. degrees from Wroclaw University of Technology, in Poland in 1978 and 1981, respectively. He received his D.Sc. degree in Computer Science from the Polish Academy of Sciences in Warsaw in 2007. He was a Visiting Professor of Computer Science at the Bialystok University of Technology, where he is now working as a Full Professor. He was with AGH University of Science and Technology from 2008–2014. He is also working as a Professor with the Faculty of Mathematics and Information Sciences at Warsaw University of Technology. His areas of interest are biometrics, image analysis and processing and computer information systems. He has published more than 220 papers in journals and conference proceedings, and edited 28 books, including 10 text- and reference books. He supervised more than 130 M.Sc. and 16 Ph.D. theses and given more than 40 invited lectures and keynotes in Europe, China, India, South Korea and Japan. He has received more than 20 academic awards and is a member of the editorial boards of over 20 international journals and conferences. He is an IEEE Senior Member and was selected as IEEE Distinguished Speaker for 2011–2016. Khalid Saeed is the Editor-in-Chief of International Journal of Biometrics, published by Inderscience.

Nabendu Chaki is a Professor at the Department of Computer Science & Engineering, University of Calcutta, Kolkata, India. He first graduated in Physics from the legendary Presidency College in Kolkata and then in Computer Science & Engineering from the University of Calcutta. He completed his Ph.D. from Jadavpur University, India in 2000. He shares 6 international patents, including 4 U.S. patents, with his students. Prof. Chaki has been active in developing international standards for software engineering and cloud computing as a member of the Global Directory (GD) for ISO-IEC. As well as editing more than 25 book volumes, Nabendu has authored 6 text- and research books and has published over 150 Scopus indexed research papers in journals and at international conferences. His areas of research interest include distributed systems, image processing and software engineering. Prof. Chaki also served as a Researcher in the Ph.D. program in Software Engineering at the U.S. Naval Postgraduate School, Monterey, CA. He is a visiting faculty member for numerous Universities in India and abroad. In addition to serving on the editorial board for several international journals, he has also been on the committees of over 50 international conferences. Prof. Chaki is the founder Chair of ACM Professional Chapter in Kolkata.

Part I
Security Systems

Enhancing Security in ROS

Gianluca Caiazza, Ruffin White and Agostino Cortesi

Abstract In recent years, we observed a growth of cybersecurity threats, especially due to the ubiquitous of connected and autonomous devices commonly defined as Internet of things (IoT). These devices, designed to handle basic operations, commonly lack security measurements. In this paper, we want to tackle how we could, by design, apply static and dynamic security solutions for those devices and define security measurements without degrading overall the performance.

Keywords Security · IoT · x.509 certificate · Encryption

1 Introduction

With the spread of connected and smart devices, we observed a tremendous grown on the amount of personal data that are stored and processed every day. Considering the sensitive nature of this information, there's a widespread suspicion concerning the way in which this information flows into the infrastructures. Additionally, with the increase of smart cities and connected environments this critically is going to enlarge. In particular, in these environments we can identify two possible group of threats: physical and logical. In the first group, we found simple physical attack as shooting down the device or capture it with a net, as well as more complex one as radio sniffing, tampering, dossing, etc. In the authors' opinion, this kind of vulnerability should be addressed from the hardware/firmware point of view, since it will result in a waste of effort tackle them from the application level. Indeed, our focus is on the logical approach of the problem, on the data-centric analysis of the infrastructure. So, in the light of this, it is easy to understand the importance of supporting the key property of computer security: confidentiality, integrity, authenticity, non-reputation,

G. Caiazza · A. Cortesi (✉)
Ca' Foscari University, Venice, Italy
e-mail: cortesi@unive.it

R. White
UC San Diego, San Diego, USA

© Springer Nature Singapore Pte Ltd. 2019
R. Chaki et al. (eds.), *Advanced Computing and Systems for Security*,
Advances in Intelligent Systems and Computing 883,
https://doi.org/10.1007/978-981-13-3702-4_1

availability. By enforcing these simple concepts, we are able to develop countermea-sures against literature attacks such as eavesdropping, modification, impersonation, repetition. However, since we are working with IoT devices, the way in which these properties are enforced is not trivial. In fact, we need to consider the intrinsic limita-tions of the devices, either from the power consumption point of view and the actual computational power available. Additionally, since we want to design a solution that could be applied to a wider scope of devices, it is safe to assume that we want to keep as real-time performance as possible.

2   Security Enhancements

In order to develop our solution, we worked on a widespread open-source middleware software for robotic implementation: Robot Operating System (ROS) [1]. It provides the services of an operating system, including hardware abstraction, low-level device control, implementation of commonly used functionality, message passing between processes and package management. Still, from our point of view, it is important to notice that it doesn't implement any security measurements at all.

Intuitively, the easiest way to ensure secrecy between two (or more) agents is by encryption. Therefore, by means of TLS/SSL and the usage of x.509 certificate we can easily enforce confidentiality and authenticity. In the particular case of ROS, this improvement has been implemented with SROS [2, 3] a set of security enhancements that aims to secure ROS. By leveraging on SROS, our goal is to extend the new security features and improve them with some fine statical mechanisms: (1) define an exhaustive standard for security logging the API operations; (2) define a new profile syntax standard for the definition of policy file; (3) provide new method for the automatic generation of the aforementioned certificate. Still, been aware of the limited resources on the devices we need to carefully evaluate each choice in order to keep the solution as lightweight as possible. Along with this line, we design our enhancements as static offline mechanisms which results are simply applied to the devices. In detail, apart from the logging mechanisms—that adds a negligible overhead on the device—the other solutions aim to granularly define access control for the agents. In addition, we discuss two different approaches for the definition of the relations between the certificates and policy profiles.

Our goal is to keep the resource usage under control. In detail, by leveraging on the statical approach defined by SROS (that embeds the access control policy as extensions of the x.509 certificate), and by exploiting Park et al. work on x.509 extensions [4, 5], we propose two different architectures for IoT network as well: user pull and server pull.

By combining our proposed static improvements with the usage of smart certifi-cates, we can easily enhance the way in which users define how agents communicate in the network.

Contrary to all the related works, to the best of our knowledge this is the first research that focuses on the automatic definition of embedded policy profiles in a trusted network and actively prevents, at the application level, the disclosure of sensitive information and blocks unauthorized agents by applying a priori access control model in addition to library functions security enhancements.

3 Technical Overview

In order to better understand how we develop the proposed solution, it is important to have a grasp of the framework that has been selected. In this section, we will briefly evaluate ROS and the general concept behind SROS.

3.1 *Robot Operating System*

ROS implements a peer-to-peer network, namely 'graph', in which the processes (agents) can communicate at run-time via publish/subscribe [6] pattern. From our point of view, the graph itself is the key concept that we need to exploit. The basic computation graph components of ROS are nodes, master, parameter server, messages, services, and topics, all of which provide data to the graph in different ways. Let's see in detail each component:

- Nodes: Nodes are the basic processes that perform computation. ROS is designed to be modular at a fine-grained scale; usually, a robot control system comprises many nodes. For example, one node controls a camera, one node controls the temperature sensors, and so on.
- Master: The ROS master provides name registration and lookup to the rest of the computation graph. In practice, the master is a DNS server for nodes. Without the master, nodes would not be able to find each other, exchange messages, or invoke services.
- Parameter Server: The parameter server allows data to be stored by key in a central location. Even though we need to consider this as an independent component, it is part of the *Master*.
- Messages: Nodes communicate with each other by means of messages. A message is simply a data structure, comprising typed fields.
- Topics: Messages are routed via a transport system with publish/subscribe mechanism. A node sends out a message by publishing it to a given topic. The topic is a name that is used to identify the content of the message. A node that is interested in a certain kind of data will subscribe to the appropriate topic. It is important to notice that there may be multiple concurrent publishers and subscribers for a single topic, and a single node may publish and/or subscribe to multiple topics. Furthermore, publishers and subscribers are not aware of each other's existence.
- Services: The topics publish/subscribe model is a very flexible communication paradigm, but its many-to-many, for *one-way* transport it's more appropriate

a request/reply interactions. This kind of communications is done via services, which are defined by a pair of message structures: one for the request and one for the reply. A providing node offers a service under a name, and a client uses the service by sending the request message and awaiting the reply.

That said, we can easily translate the ROS's structure into a more general IoT network, in which the nodes represent the single device and the master the default gateway; indeed, this is the structure that is currently in the market (e.g. Apple HomeKit, Samsung Smart-home, Google Things Solutions).

3.2 SROS: Secure ROS

As mention, SROS is a set of security enhancements for ROS that aims to secure ROS API and ecosystem by means of native TLS/SSL support for all IP/socket-level communication. In addition, with the usage of x.509 certificates, they defined chains of trust by means of a certificate authority (namely keyserver), namespace node restrictions and permitted roles, as well as user-space tooling to auto generate node key pairs, audit ROS networks, and construct/train access control policies. Furthermore, they defined AppArmor profile library templates that allow users to harden or quarantine ROS-based processes running at Linux OS kernel level.

That said, we can summarize that SROS is intended to secure ROS across three main fronts:

1. Transport Encryption: with the usage of TLS and x.509 PKI for authenticity and integrity
2. Access Control: restrict node's scope of access within the ROS graph to only what is necessary leveraging on definable namespace globbing
3. Process Profiles: restrict the application (file, device, signal, and networking access) thanks to AppArmor profile component library for ROS.

4 Access Control Policy Generation

The access control policy profile should not be confused with that of SROS's AppArmor profile library or the profiles they provide using features from Linux security modules. Those help users provide Mandatory Accesses Control (MAC) for ROS nodes on the run-time processes level of the hosting operating system. The policy profiles we are discussing below are the ones related to enabling access control for ROS nodes in the ROS graph network level.

Generally speaking, in these profiles we store the communication topology between agents, specifying the operation that is allowed or denied to the device. In order to build a proper profile, we can proceed in two different ways: manually defined the rules or automatically extract them from meaningful log.

As representative of the second group, we have the aforementioned `AppArmor`, a proactively software layer that protects the operating system and applications from external or internal threats, by enforcing good behaviour and preventing even unknown application flaws from being exploited. In detail, AppArmor defines a set of rules for the selected application based on the operation that has been *exercised* by the user during a training phase.

4.1 Security Logging

Along the same line, we want to implement a similar mechanism also in our solution. First of all, we need to specify a suitable way of acquiring security logs. In order to do so, we need to identify the key components and communication mechanisms (e.g. API) that are used in the application. The new logging system that we are going to define will be more structured and informative in regard to the required operation and resource access.

Therefore, leveraging on the well-known Unix logging system, we specify three different execution modes: *audit*, *complain*, and *enforce*. In the first mode, we log all the operation that are executed by the application without applying any constraint. Once the user has exercised the application as necessary, he can move to complain mode. In this case, we prompt to the log all the operations that violate the rules that has been defined in the previous mode. This is a crucial phase, in which we can verify if the rules that we previously define works as expecting or are too strict or naive with respect to the desired access control. Then, if the defined profile works as wanted, it will be enough to enforce them by applying the namesake mode.

All in all, in this phase, is important to enhance the behaviour of the application's APIs such that we can easily log the kind of resource that the agent wants to access. In the case of ROS, we have extract three macro-group of resources: topic, service and slave operation. In detail, in the topic we found the APIs that allow a node to register himself as a publisher or as a subscriber. Along the same line, we have the service APIs that operates likewise the topic once; lastly, we have all the operation that are executed by simple slave node. In general, we can make a parallelism with this API and the hierarchy in an IoT network such that slave operations define the IoT device API, whilst topic and service represent the communication mechanisms (gateway API).

4.2 Policy Profile Syntax

As previously said, we took a good deal from the AppArmor policy definition such as it's globbing syntax. Our goal is to define a more applicable syntax intended to encode policy rules, definition, and relationships in our trusted network. Our intuition is to leverage upon the namespace resource organization a good deal to define the profiles.

In ROS, we define a domain as a simple root '\'. In it, we address a node simply recalling its position in the hierarchy. When nodes are integrated into a larger system, they can be pushed down into a suitable namespace that defines their functionality. For example, one could take two robots namely `foo` and `bar` and merge them into the same domain with 'foo' and 'bar' subgraphs. Therefore, if both devices had a node named '*camera*', they would not conflict since they will be addressed respectively as \foo\camera and \bar\camera.

It's important to notice that ROS supports several methods to address a resource that could be either *base*, *relative*, *global* and *private*; although for the sake of our policy profile, we always extend them in a plain version that explicit all the chain relation of the node. In this way, we can define policy profile in an agnostic way in regard to the underlying implementation.

Since we are able to unambiguously address an agent, we can specify the necessary rules for each one. We can suppose that several devices share a set of common rules, either because they are necessary for a feature (e.g. logging), or because the node has a *role* in the application, that is an administrative device. In order to simplify the management of these roles, we introduce the concept of `include`. As for other programming languages, the goal of using include statement is to add—at compile time—a set of predefined piece of code, in this case a collection of rules for the specific device. These rules simply follow the same structure of the one that are defined in the policy profile.

Intuitively, we can sort out different resource twofold, by explicitly define the kind of resource (i.e. topic, service, parameter) and by defining resource specific masks for permissions. Let see below an example of node profile:

```
/namespace
{
#include role
resource /scope masks
}
```

As mention before, one of the advantages of using this kind of notation for addressing resource and agent is the usage of globbing syntax. In detail, it's possible to define regexp formulas in rules and profiles scope. Therefore, if we want to specify that an agent is allowed to interact with all the first-level cameras of the other agents, it will be enough to specify the rule: **\camera.

As a rule of thumb, we define regular expressions with the following syntax:

- * : It represents any number and any characters in the current namespace.
- ** : It represents any number and any characters including the definition of sub-namespaces.
- ? : It represents a single character or number.
- [abc] : It represents the single character a, b, or c.

- [a–c] : It represents all the character in between a and c.
- {ab, cd} : It expands the string to match the expressions ab and cd.

Additionally, we introduce the possibility of specifying deny rules. In fact, if we want to single out a resource from a wider regexp, it will be enough to define a specific rules for the resource (or resources via regexp) as shown below:

```
....
deny resource /bar/foo1 masks,
resource /bar/foo* masks,
....
```

5 X.509 Certificate: Distribution Architecture

In this section, we evaluate different approaches for the definition of the relations between the certificates and policy profiles. We discuss two different architectures: user pull and server pull.

As the name suggests, these models are, respectively, user based and host based. In particular, the purpose of these architectures is to discuss the static and dynamic solution either for certificate distribution and attribute verification.

5.1 User-Pull Architecture

As the name suggests, in this mode the user pulls the attributes certificate bundle from the server and stores it locally. Then, it uses the certificate for the authentication phase with the other agents in the graph. This family of solutions exploits the problem of the authentication leveraging on the integrity services offered by the certificates by design. In fact, the certificates are issued by the certificate authorities (CA) which are trusted entities in the system. However, in addition to the trivial implementation, in which we store the profiles as extensions of the x.509 certificate, we want to introduce additional types of certificate that are defined in the x.509 standard: *Identity certificates* (ID) and *Attributes certificates* (AC) [7].

While X.509 public-key certificates bind an identity and a public key, an attribute certificate (AC) doesn't contain a public key but may contain attributes that specify group membership, role, security clearance, or other authorization information associated with the AC holder. The reason why we should use AC for authentication information in place of PKC extension is mainly for two reasons: authorization often does not have the same lifetime as the identity and the public key. So, when we store authorization information in a PKC extension, the general result is that we are shortening the PKC useful lifetime. Furthermore, the CA that issues the PKC is not usually authoritative for the authorization information. In fact, it represents a threat for the system that should be avoided. There are several ways in which we can bound the authorization with the identity certificate; below we present three different approaches: *monolithic, autonomic*, and *chained signatures*.

Monolithic

This is the easiest solution in which we consider only one certificate authority in charge of both identity certificate and attribute. In the monolithic approach, we create a certificate that holds both the identity and attributes information; trivially, this is implemented by using x.509 and the extension fields. The resulting certificate *tightly coupled* the identity information and the attributes with a single signature. This means that in order to change an information in an existing certificate, we need to revoke the previous and issue a new updated one. However, the management of this particular solution is simplified in comparison with the other, since we need to trust only one CA. Considering that all the certificate information are verified by the only CA's signature, and therefore, there is only one CRL that needs to be checked.

However, as previously said, this approach has several drawbacks. First, multiple CAs are not supported. We can't revoke a certificate if we aren't the issuer CA and there is the possibility of issuing multiple certificates with different attributes for the same agent. Secondly, due to the design of the solution we are not able to maintain different lifetime for multiple attributes; in fact, all the attributes share the lifetime of the PKC. All in all, monolithic approach is the most favourable solution when we are limited in terms of resources or we are looking for a statical solution, even though we sacrifice flexibility in maintenance.

Autonomic

In this case, we introduce the concepts of multiple CAs and we differentiate between identity and attribute certificates. With this approach, we want to define a *loosely coupled* binding between the ID certificate and the AC. This particular solution allows the existence of multiple ID certificates per agent provided that there is an injective function from the certificates to the agent; this means that we will never have more than one agent that corresponds to an ID certificate. As such, we can bind each AC with a different set of information from the ID such as subject's name, public key, certificate serial number (Fig. 1).

Depending on the chosen set of information that has been selected, we can modify the certificate issuing a new one and still maintain the correlation between the AC and ID as long as the binder information has not been changed. As an example, if we insert in the ID certificate the unique serial number of the agent and we bound the attribute certificate based on that, we can change the other information such as lifetime, serial number, subject's name, while the link between the certificates holds. However, since we moved from a static solution to a dynamic one, we should be extra careful about the new threats as the choice of the information set. In fact, even though we have an injective function from the certificates to the agent there aren't constraint on the information that are stored in the certificates. It means that if we accidentally choose a common set of information, it may happen that the same attribute certificate can be used by unauthorized agents that share the set of information with the authorized agent.

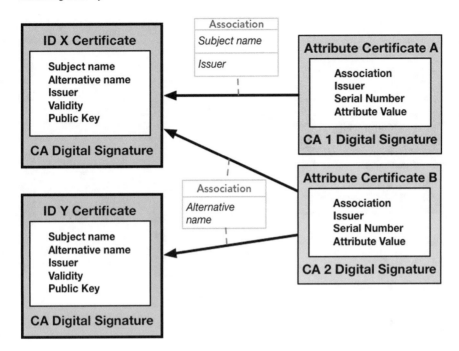

Fig. 1 Autonomic certificate

Chained Signature

With this technique, we want to take the security features guaranteed by the monolithic solution and part of the flexibility of the autonomic approach and create a new hybrid solution. As in autonomic, an agent can have multiple ID certificates issued by multiple CAs. However, instead of binding the attribute certificate with an arbitrary set of information, we bound on the digital signature of the corresponding ID certificate. In fact, if the information in the referenced ID certificate is changed (a new certificate is issued), the digital signature *should* change as well [8]. Under the assumption that we are using a suitable signature algorithm, when we issue a new ID certificate (with a different signature) the link between the two certificates is broken and then the attribute certificate becomes automatically useless. One of the advantages of chained signatures is that we don't need to aggregate all the attributes according to the shortest lifetime of the certificate as in monolithic. Furthermore, we introduce a mechanism that allows us to share with other agents only the necessarily information. In fact, with a monolithic certificate all the policy of the agent are available in the PKC, instead of this solution we can share only the necessarily attribute certificate enforcing a new privacy feature and dynamic management of profiles (Fig. 2).

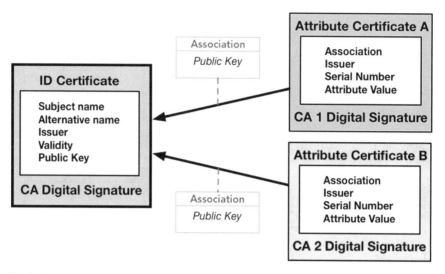

Fig. 2 Chained certificate

5.2 Server-Pull Architecture

In server-pull model, the general idea is to demand the authentication phase to the attribute authority (AA). Our goal is to define a dynamic solution in which apart from the AA, no one needs to know the attribute information. However, as in user-pull model, we still need a method to unambiguously identify an agent in the graph; we can achieve this straightforwardly with the usage of the already defined ID certificates.

But instead of issuing attribute certificates and bound them to the ID, we store the attribute policy roles locally in the attribute authority. This particular solution allows us to implement in the AA whenever access control we want (e.g. MAC, DAC, MLS, MCS, RBAC), without modifications to the client (agents) APIs. We design a high-level API that permits the agents to retrieve the authorization response from the AA regardless of the chosen access control method.

There are several advantages on this model: first of all, thanks to the usage of AA instead of static certificate we can achieve a dynamic flexible solution that can evolve and change during run-time without additional setup. Secondly, we can maintain the secrecy of the sensitive information about the policy topology inside the attribute authority without compromising the network topology in critical applications.

Still, the solution introduces the problem of the *single point of failure* (SPOF). In practice, considering that all the agents need to query the attribute authority in order to receive a response about the permissions, if it fails the entire system will stop working. In fact, from the attacker point of view it will be enough to tamper the AA to compromise the entire system. In addition, we introduce by design an overhead in the handshaking process; in fact, what was a straightforward local check of the profile in user model became a remote request to the external attribute authority.

However, this problem can be slightly mitigated by resuming previous sessions via `Session Ticket` as introduced in TLS 1.3 [9].

6 Related Work

To the best of our knowledge, the present work is the first research focusing on the discussion of a set of tools and techniques for the automatic definition of security policy profiles in a robotic framework. Although several security threats analysis in the industrial robotic applications have been performed recently, they mainly focus on the threats deriving from the modification of network topology from local to remote connection [10–12].

Akerberg et al. [13] tackled the problem from the communication channel point of view, proposing a security communication framework for the integration in classical wired industrial networks of wireless nodes, with the implementation of end-to-end integrity and authentication measures by utilizing the black channel concept. On the other hand, Wang et al. [14] analysed publish/subscribe communication paradigm in a wide-area network. In this particular setup, services handle information across distinct authoritative domains and need to manage a large population of publishers and subscribers. In detail, they discuss about the security issues and requirements that arise distinguishing among those requirements that can be achieved with current technology and those that require novel solutions.

Similarly to the proposed solution, Dieber et al. [15] proposed a security architecture intended for use on top of ROS on the application level. By means of an authorization server and the usage of x.509 certificates for authentication, they ensure that only valid registered nodes are authorized to execute some operations. However, their statical architecture is based on the assumption that we have manually generated and distributed the certificates and registered the list of nodes in the authorization server; moreover, they delegate to the user the distribution of the lists of certificates serial number of the nodes that are authorized to query each other.

Additionally, Lera et al. [16] presented an interesting analysis on which they proposed that ROS communications should be encrypted. Differently to the previous discussions, in this document they directly evaluate the performance of ROS under encryption. In detail, they used 3DES cyphering algorithm and evaluated the performance both from the computing and the communications' point of view. These works are particularly useful in evaluating the performance of ROS under encryption, although it's important to notice that we should prefer algorithms that have dedicated instructions added in hardware in modern CPUs, that help them run substantially faster than software only ciphers implementation. So, depending on the chosen algorithm and key length we can easily observe some changes in the performance [17, 18].

7 Conclusion

We have presented a static procedure to generate and amend policy profiles for IoT robotic devices. In order to determine a suitable solution, we analysed the ROS framework and we discussed about the design of an agnostic solution for the definition and application of an access control policy profile. In the first part, we discussed about the definition of a new standard for security logs that allows us to reconstruct all the library functions call-back that are not covered by the 'basic' log system, by means of a more accurate and defined structure. Then, we proposed a standard syntax for the definition of the policy profiles based on the well-known AppArmor syntax. Finally, we discussed about static and dynamic solution for certificate and attribute distribution, thus contributing to the scenario depicted in [19]. From our analysis, it emerges that one of the biggest threats of robotic network is the lacking of security measure in the communication mechanisms that needs to be harden with the introduction of access control and encryption mechanisms.

Still, there are a number of issues that are part of our plans for future work as mitigate the disclosure of sensitive information (i.e. an agent profile), improve privacy in access to partially unauthorized resources (e.g. function output custom sanitization), as well as decoupling the cryptographic operations from the authenticity mechanisms by means of middleware implementations as oneM2M [20] or other DDS.

Moreover, we aim to conduct a further analysis in the definition of a real-time system with the addition of cryptographic mechanisms via 'Real-Time Publish Subscribe' (RTPS) protocol for mission-critical implementations.

All in all, we believe that we have given a solid base for the definition of the future security mechanisms for robotic devices that could be easily and securely integrated in big-scale deployments without suffering software limitations. Furthermore, in our opinion the definition of high-level solutions as the one that has been proposed in this paper is critical for spread security solutions by reducing the tradeoff between security and usability.

Acknowledgements Work partially supported by CINI Cybersecurity National Laboratory within the project FilieraSicura and by the Executive Program 2017–2019 Italia-India within the project 'Formal Specification for Secured Software Systems'.

References

1. Quigley, M., Gerkey, B., Conley, K., Faust, J., Foote, T., Leibs, J., Berger, E., Wheeler, R., Ng, A.: ROS: an open-source robot operating system. In: ICRA Workshop on Open Source Software (2009)
2. Quigley, M., White, R., Christensen, H.I.: SROS—Securing ROS over the wire, in the graph, and through the kernel. ROSCon (2016)

3. White, R., Caiazza, G., Christensen, H., Cortesi, A.: SROS1: using and developing secure ROS1 system. In: Robot Operating System (ROS): The Complete Reference, vol. 3. Springer (2018) (to appear)
4. Park, J.S., Sandhu, R.: Smart certificates extending X.509 for secure attribute services on the web. In: Proceedings of the of 22nd National Information Systems Security Conference (NISSC), pp. 337–348 (1999)
5. Park, J.S., Sandhu, R.: Binding identities and attributes using digitally signed certificates. In: Proceeding ACSAC 00 Proceedings of the 16th Annual Computer Security Applications Conference, p. 120 (2000)
6. Eugster, P.T., Felber, P.A., Guerraoui, R., Kermarrec, A.M.: The many faces of publish/subscribe. J. ACM Comput. Surv. (CSUR) 35(2), 114–213 (2003)
7. Farell, S., Housley, R., Turner, S.: An Internet Attribute Certificate Profile for Authorization. Internet Engineering Task Force (IETF) (2010)
8. Lenstra, A., Wang, X., de Weger, B.: Colliding X.509 Certificates, Report EPFL (2005)
9. The Transport Layer Security (TLS) Protocol Version 1.3. https://tools.ietf.org/html/draft-ietf-tls-tls13-18
10. Cheminod, M., Durante, L., Valenzano, A.: Review of security issues in industrial networks. IEEE Trans. Ind. Inform. 9, 1 (2013)
11. Byres, E., Dr, P.E., Hoffman, D.: The myths and facts behind cyber security risks for industrial control systems. In: Proceedings of VDE Kongress (2004)
12. Dzung, D., Naedele, M., von Hoff, T., Crevatin, M.: Security for industrial communication systems. Proc. IEEE 93(6), 1152–1177 (2005)
13. Akerberg, J., Gidlund, M., Lennvall, T., Neander, J., Bjorkman, M.: Efficient integration of secure and safety critical industrial wireless sensor networks. EURASIP J. Wirel. Commun. Netw. 1, 1–13 (2011)
14. Wang, C., Carzaniga, A., Evans, D., Wolf, A.: Security issues and requirements for internet-scale publish-subscribe systems. In: Proceedings of the 35th Annual Hawaii International Conference on System Sciences, 2002. HICSS 2002, pp. 3940–3947
15. Dieber, B., Kacianka, S., Rass, S., Schartner, P.: Application-level security for ROS-based applications. In: Proceedings of 2016 IEEE/RSJ International Conference on Intelligent Robots and Systems (IROS) (2016)
16. Lera, F.J.R., Balsa, J., Casado, F., Fernandez, C., Rico, F.M., Matellan, V.: Cybersecurity in autonomous systems: evaluating the performance of hardening ROS. In: XVII Workshop en Agentes Fsicos (2016)
17. Singh, G., Supriya, A.: A study of encryption algorithms (RSA, DES, 3DES and AES) for information security. Int. J. Comput. Appl. 67(19), 09758887 (2013)
18. Giry, D.: Bluecrypt cryptographic key length recommendation. http://www.keylength.com/. Accessed Oct 2016
19. Cortesi, A., Ferrara, P., Chaki, N.: Static analysis techniques for robotics software verification. ISR 2013: 1–6
20. Datta, S.K., da Costa, R.P.F., Bonnet, C., Harri, J.: oneM2M architecture based IoT framework for mobile crowd sensing in smart cities. In: Networks and Communications (EuCNC) (2016)

Byte Label Malware Classification Using Image Entropy

Ayan Dey, Sukriti Bhattacharya and Nabendu Chaki

Abstract Malware continues to be an ongoing threat to modern computing. In our research, we present a byte level malware classification technique which is basically an improvement on an existing work [1]. We introduced an information theoretic point of view an already existing image-based malware detection method. The introducing entropy filter helps to identify the hidden patterns introduced by certain packers and encryptors, hence yields better accuracy and false positive rate than the existing method. We have implemented a proof-of-concept version of the proposed technique and evaluated it over a fairly large set of malware samples cover different malware classes from different malware families and malware authors.

Keywords Malware classification · GIST · Entropy · Byte level analysis

1 Introduction

Malware is a global outbreak, the number of new malware and their variants are growing exponentially in the World Wide Web to gain access to remote computer networks or for monetary profits. Innumerable interpretations have been offered to represent a malware. For instance, McGraw and Morrisett [2] define malicious code as "any code added, changed, or removed from a software system in order to intentionally cause harm or subvert the intended function of the system". According

A. Dey (✉)
A. K. Choudhury School of Information Technology, University of Calcutta,
Kolkata, India
e-mail: adakc_rs@caluniv.ac.in; deyayan9@gmail.com

S. Bhattacharya
Environmental Informatics, Dept. of Environmental Research and Innovation (ERIN),
Luxembourg Institute of Science and Technology, Esch-sur-Alzette, Luxembourg
e-mail: sukriti.bhattacharya@list.lu

N. Chaki
Department of Computer Science & Engineering, University of Calcutta, Kolkata, India
e-mail: nabendu@ieee.org

© Springer Nature Singapore Pte Ltd. 2019
R. Chaki et al. (eds.), *Advanced Computing and Systems for Security*,
Advances in Intelligent Systems and Computing 883,
https://doi.org/10.1007/978-981-13-3702-4_2

to Christodorescu et al. [3], malware is a program whose objective is malevolent. Vasudevan and Yerraballi [4] define malware as a universal term that encircles virus, trojan, spyware and more invasive codes.

Often, malware writers have a keen interest to generate a new variant of malware using some automated tools (e.g., metamorphic engine). Such tools have a tendency to sneak hidden malicious commands into an executable file using the same modules, to generate new malware variants from the same class. The different variants generated by the tool can employ first-order obfuscation techniques that disguise the malicious commands so that they pass through the traditional signature-based detection of anti-virus programs. But, leaves suspicious patterns in the bit level, that can be identified by means of the measure of uncertainty. In 1948, Claude Shannon [5] proposed entropy as the measure of uncertainty that can be used to summarize feature distributions in a single number.

We propose an entropy-based byte level analysis technique to classify malware. The proposed work is an enhancement over the image- based malware detection approach proposed in [1]. We have detailed this in Sect. 4.

In this paper, we introduced an entropy filter before estimating the GIST feature on the gray-level image (as shown in Fig. 2). The entropy filter does the following, (a) estimates local entropy on a pre-defined neighborhood, (b) quantifies the "amount of structure" in the entropy signal, (c) finds the position of mean change points and computes the energy distribution over multiple spatial scales. As a result, the relative change in entropy is measured. A significant change in entropy could differentiate a malware into diverse classes. Based on an empirical study, our proposed method can effectively categorize malware with 98.07% accuracy and 0.019% false positive rate. In both cases, our approach shows better performance compared to Nataraj et al. [1].

The remainder of the paper is organized in the following way. Section 2 describes the background of current state of art of malware analysis and image processing in detail. A detailed analysis of related work is given in Sect. 3, mainly highlighting the byte level malware analysis methods. We depict the detailed architecture of our malware classification method in Sect. 4. The empirical results are shown in Sect. 5. Finally, we conclude the paper in Sect. 6.

2 Background

Malware analysis algorithms that will detect and classify malware can be primarily categorized as either static or dynamic analysis. Several static analysis methods such as control flow graph based [6], instruction based [7, 8], similarity-based analysis, and byte level [9] analysis. However static analysis fails to detect malware for packed executable. Similarly, the dynamic analysis also suffers from execution overhead. The hybrid analysis combines the prospects from both static and dynamic analysis to get rid of these above issues. Another malware analysis technique, namely manual reverse engineering which is very time-consuming. However could be used when static, dynamic and hybrid analysis fails. In the following section, we would point out some preliminary malware analysis techniques in brief along with byte level content analysis technique.

2.1 Malware Analysis Techniques

Analyzing malware comprises a collection of tasks, some simpler than others. This endeavor can be assembled into stages that grow upwards in complexity based on the characteristics of the associated malware analysis techniques as shown in Fig. 1.

Most often, when performing malware analysis, it is impractical to infer the source code's availability for every program [10]. The only preferable solution left is executables with certain juridical bindings which won't be human-readable. The information from the executable can be accumulated either by executing them or operating static reverse engineering or by using both methods.

Analysing Byte Level Content This basic static malware analysis technique does not require specialized knowledge of reverse engineering. Different statistical and information-theoretic feature extraction techniques are used to extract features from the malicious executables without disassembling or decompiling it. These techniques

Fig. 1 Malware analysis techniques

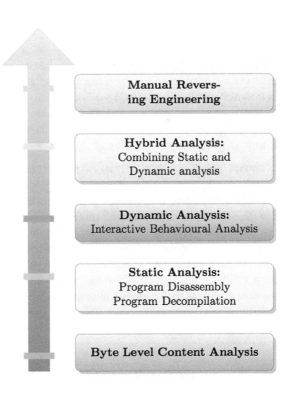

speculate what is the next pattern of a set of data would be or extract the principal components of a set of data and determine the common pattern among them. It views binaries simultaneously from different structural perspectives and performs statistical analysis or machine learning techniques on these structural fingerprints.

Static Analysis Having a specialized knowledge of disassembly, code constructs, operating system etc. Advanced static analysis having a specialized knowledge on disassembly, code constructs etc., and can generate a steeper learning curve than the ordinary static analysis [9]. A static analysis can globally view an entire executable, considering all executable paths and can identify the logic of an executable without executing it. Some algorithms are applied to generate a transitional representation of it using reverse engineering which includes disassembling and decompiling. The major weakness of static analysis is that self-modifying programs [11] and packed executable is out of its scope.

Dynamic Analysis Sometimes static analysis is not enough to detect the properties of a malware. Some automated static analyses tools may fail to produce satisfactory results during the investigation. Due to this fact, an analyst might decide to make a closer look at the exhibits nature. Behavioral analysis is the "quick and dirty" way of malware analysis. Dynamic analysis needs the code execution and monitoring its execution path in an authentic or a virtual system. The prospects of such analysis come from the fact that malware executes its own pre-defined way as soon as it is initiated and when it is being executed, therefore makes it robust to Code Obfuscation. Dynamic analysis provides the control and data flow information although it suffers from the execution overhead.

Hybrid Analysis The hybrid analysis integrates the static and dynamic both methods to form the detection and classification task more accurate and robust to modification in malware progression. Hybrid methods mainly analyze the file prior to execution. This static–dynamic combination tries to construct a model through which a malware analyst is able to understand and instrument the code. Although the hybrid analysis is a complex technique for obfuscated programs.

Manual Code Reversing The manual analysis is required when circumstances prevent the application of automated reverse engineering techniques. A malicious code having some malicious specimen could summarize some fruitful insights with the available decision, during manual reverse engineering technique. A simple reverse engineering tool composed of a disassembler and a debugger could be aided by some plugins and a compiler that automate some aspects of this effort. IDA Pro[1] is probably the most popular disassembler used for reverse engineering of malware. The primary constraint of this method are, firstly, it takes so much time and secondly, it requires a skill set that is comparatively unique.

---

[1]https://www.hex-rays.com/products/ida/.

2.2 Texture Analysis of an Executable

Feature Vector Each malware variants from a specific class exhibits some significant amount of textures. This information could be very helpful for malware researchers. Several features (e.g., Gabor filter) may be considered to analyze textures. In this paper, a Gabor like a feature is used to classify malware. We use GIST [12, 13] to compute texture features, using the wavelet decomposition technique. Each image location is expressed by a set of filter outputs adjust to distinct orientations and scales. A steerable pyramid having eight orientations and four scales is applied to the image to capture the regional information. Then mean value of magnitudes of the regional features is estimated over broad spatial regions. A more comprehensive description of GIST features can be found in the literature [12]. GIST feature is mainly used for Scene and object classification.

Classification In this paper, k-nearest neighbors along Euclidean distance for classification is used. In our experiments, we have randomly chosen two-third of data for training and left one-third is considered as a testing purpose. A test sample is classified to a class if the Euclidean distance between the class and the sample is least among all.

2.3 Image Entropy in Malware Analysis

In information theory, an entropy evaluates the uncertainty of a state. In 1948, Shanon [5] defined the entropy H of a discrete random variable X, where X is a set of all possible values and $P(X)$ denotes probability mass function (Eq. 1).

$$H(x) = E[I(X)] = E[-ln(P(X))] \tag{1}$$

Entropy could be very helpful for the random distribution of probabilities. A highly random distribution shows high entropy values, whereas entropy could fall with low randomness. In malware byte samples, randomness should be very high rather ordinary files. We are trying to identify that randomness using local entropy filter. A local entropy of a pixel suggests an entropy value in an N-by-N area around that pixel. In our algorithm, we have considered the value of N is 9. Basically, this local entropy helps to calculate the statistics for that neighborhood. For example, if the neighborhood values are similar to each other then it will produce a lower entropy and vice-versa.

3   Related Work

There are several malware detection and classifications methods. Some of these are graph based [6, 14], instruction sequence based [7, 8], instruction frequency based [15], API call based [16–18] and behavioral based [19–21]. In this paper, we limit our focus only on byte level content-based malware analysis.

In 1994, Kephart [22] presented a method to automatically extract signatures for the malware using n-gram for the first time. In [23], Schultz et al. used diverse data mining methods to differentiate between the benign and malicious executable in Windows or MS-DOS format. Three different methods are used, to statically extract features from executables. The first technique excerpts DLL information from the portable executables. The second technique extracts strings (as a feature), from the executables using GNU strings program. The third feature extraction technique uses byte sequences (n-grams) using hexdump. Memory consumption was a scalability bottleneck. In 2006, Kolter et al. use n-gram analysis and data mining techniques to detect malicious executable in the wild. Top n-grams with the highest entropy are considered as binary features for every PE file. The authors did a detailed study to determine the size of n-grams as 4-grams, the size of words as one byte and the number of top n-grams as 500 n-grams. Finally, they classify their technique for two classification issue: (1) classification between the benign and malicious executables, and (2) categorization of executables as a function of their payload. The authors categorize only three types: mailer, backdoor, and virus due to the finite number of malware samples. They experimented on two small datasets, one of 476 malware, 561 benign-ware (95% accuracy with 5% FP in validation); the second of 1971 benign-ware, 1651 malware (94% accuracy and 1% FP in validation) [24]. in 2007, Robert Lyda et al. showed that binary files with a higher entropy score tend to be correlated with the presence of encryption or compression. Applying this fact, the entropy score of memory space is continuously changed while packed instructions are unpacked into memory. They compared more than 20,000 malware to check whether they are able to detect these concealment methods. However, they did not consider malware detection [25]. In the same year, Stolfo et al. used n-gram analysis for labeling the different file types [26] and later for malware detection [27]. File print analysis [26], uses 1-gram byte distribution of an entire file and compares it with diverse file type models to understand the type of the file. Whereas in [27], they detect malware embedded in DOC and PDF files using three distinct models, single centroid, multi-centroid, and exemplary of benign byte distribution of the whole files. However, their proposed technique is specific to embedded malware and does not deal with detection of stand-alone malware. In [9], authors compute a wide range of statistical and information-theoretic features to quantify the byte level file content, in a block-wise manner. The feature vector of every block is generated and is transferred to some typical data mining methods (J48 decision trees) that categorize the blocks as regular or probably malicious. The limitations of these previous n-gram-based vector models were studied in 2010 by Stibor [28]. Thomas Stibor investigated and compared machine learning algorithms on originally infected executable and malware loader files. Experimental results reveal that N-gram representations of an executable file that are infected can be hardly detected with machine learning methods. Santos et al. in 2011 introduced a semi-supervised methodology to reduce the labeling process. Their n-gram vector was the frequency of all possible n-grams, an important scalability limitation. After experiments on 1000 malware and benignware, they reported 89% of accuracy with 10% of false positives [29].

4 Malware Classification Using Image Entropy

The proposed image-based byte level malware classification approach consists of following 4 steps,

1. Byte to Image Conversion.
2. Entropy Filter.
3. Feature Extraction.
4. Classification.

Figure 2 illustrates the steps stated above.

Training: Block A is mainly responsible for entropy calculation and feature extraction purpose. "Byte to Image Converter" transformed byte samples from an executable into a grayscale image. Next, an entropy filter is used to construct the cor-

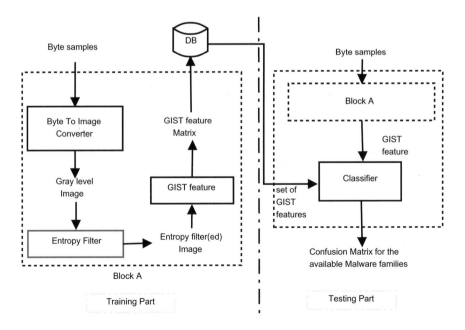

Fig. 2 Block diagram of proposed Method

responding entropy image depends on the local entropy values of the gray images. Finally, GIST feature [12] is calculated and stored in a database for malware classification.

Testing: Given a malware executable as input, passed through Block A, yields a set of GIST features to be classified using k-NN classifier.

4.1 Byte-to-Image Conversion

A byte file basically contains a one-dimensional vector of 8-bit integers. This 1D vector can be easily transformed into a two-dimensional array. This can be easily anticipated as an 8-bit grayscale image where each pixel belongs to the range [0 (black) to 255 (white)]. The width of the image is fixed. Depending on the size of that 1D vector height could vary. Based on the empirical observations, Nataraj et al. [1] provide some recommended image widths for distinct file sizes. These two-dimensional files are then transformed into gray images. Next, each transformed grayscale image contains some layout and textures that could be an interesting feature [1] [30].

4.2 Entropy Filter

Instead of grayscale image of each malware, similarity analysis is a tedious job. We construct an image that contains local entropy values of an N-by-N neighborhood to solve this issues. In our approach, each pixel of the output image contains the entropy value of the 9 by 9 neighborhood around the input pixel.

Entropy; An entropy measures the statistics of randomness. It provides the information about the local variability of intensity values of an image. For example, in an area with smooth the texture, the range of neighborhood values around a pixel should be very small. On the contrary, in an area with the rough texture, the intensity values differ heavily from one another. Thus, it can easily characterize the texture of an image. Entropy is calculated using the following Eq. 2,

$$Entropy = -\sum_{0}^{255} P_i * log_2 P_i \qquad (2)$$

In Eq. 2 P_i denotes the probability of appearance of a byte value i [31]. The range considered between 0 to 255 since the bytes values appears in the same set.

4.3 GIST Feature

As per the experimental results from the psychophysical studies, it can be clearly shown that human eye analyzes textures by decomposing the image into its frequency and orientation levels [32]. Frequency and orientation representations of Gabor filter are almost similar to human visual systems. Thus, Gabor is well suited for texture representation and discrimination. A two- dimensional Gabor function contains a sinusoidal plane of finite frequencies and orientations that are modeled by Gaussian envelope [1]. In Gabor filter, the set of frequencies and orientations are selective. A little variation on the set of frequency and orientation creates a set of Gabor filters. An image passed through this set of filters should produce a set of filtered images through which textured features are extracted. In Nataraj et al. [1], another feature, named GIST [12, 13], is used to characterize and classify malware samples. GIST feature uses wavelet decomposition of an image which is very useful for scene classification and object classification. In this paper, a pyramid with eight orientations and four scales is applied to the image. Local representation of an image is produced by the following Eq. 3,

$$V^L(x) = V^k(x)_{k=1 To N} \tag{3}$$

where N denotes the number of sub-bands and N = 20. They compute the mean of the magnitude of local features averaged over large spatial regions shown in Eq. 4 to capture global properties of an image.

$$m(x) = \sum_{x'} |v(x')| w(x' - x) \tag{4}$$

where $W(x)$ is the averaging window. This output is downsampled to the spatial resolution of M by M pixels. Nataraj et al. [1] takes M=4 that produces a feature set of size M by M by N equals 320 features. In [12], Gist is clearly explained in detail.

5 Experiments

We compare the performance of our work with Nataraj et al. [1]. We particularly observe this work since this is one of the pioneering works that used images for malware detection. Besides the concept of using image processing for malware detection remains relatively nascent. However, the early results cited in [30, 33] are quite promising. In this section, we present an empirical study on the proposed framework. A comparative performance analysis is done to establish the superiority of the proposed approach.

Table 1 Malware Dataset from 9 families

Index	Class name	Total data	Train set	Test set
1	Ramnit	1525	1038	487
2	Lollipop	2512	1678	834
3	Kelihos_ver3	2942	1962	980
4	Vundo	471	317	154
5	Simda	42	29	13
6	Tracur	741	500	241
7	Kelihos_ver1	396	266	130
8	Obfuscator.ACY	1238	835	403
9	Gatak	1055	709	346
Total	9 Classes	10,922	7334	3588

5.1 Dataset

In our experiment, we used a dataset from Microsoft Malware Classification Challenge (BIG 2015)[2] [34]. They provide a set of known malware from nine different families. Each malware is identified by an Id. This Id is a 20 character hash value that can uniquely identify the file, a class, and an integer representing one of the family names to which it actually belongs. Table 1 shows the number of malware in each of the 9 classes. Each file contains the raw data (without PE header) in the hexadecimal representation of the file's binary content. Each binary content is transformed into a grayscale image for image processing. We generated 10,922 malware images from the dataset. The dataset is split randomly into a training set of two-thirds of the data with the remaining one-third used for the testing.

5.2 Experimental Result

In Figs. 3 and 4, compares the proposed modification with [1] based on the True Positive (and False Negative) data per malware families, the vertical axis shows no of True Positive samples (in Fig. 3) and no of False Positive samples (in Fig. 4). In each case, horizontal axis indicates class index(s). Surprisingly, image entropy-based method (shown in gray line) performs better than Nataraj et al. (shown in black line) [1] in both.

---

[2]https://www.kaggle.com/c/malware-classification.

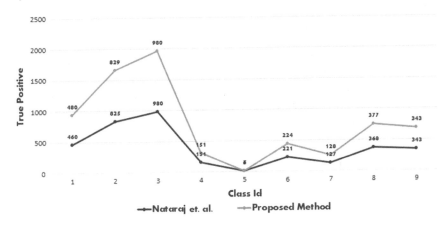

Fig. 3 True positive comparison

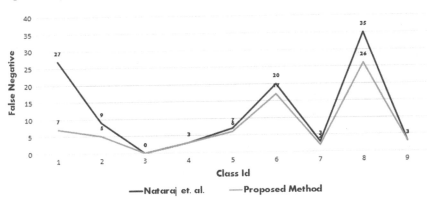

Fig. 4 False negative comparison

6 Conclusion

Image-based malware analysis has some interesting features as described in [30]. Firstly, they help malware researchers to figure out the behavior of malware, without dissection it. That will greatly reduce the execution time. Secondly, such images can be used to identify malware classes [30]. Unlike static analysis, the proposed method neither rely on the reverse engineering methods, nor attempts to resolve a (possibly over-approximated) set of all feasible behaviors by dissecting the instructions as they are found in the binary image. Unlike dynamic analysis, the proposed method does not rely on monitoring execution traces by running the malware in a controlled environment for a finite time or until the execution is accomplished. We provide experimental proof that the training set has a large impact on the resulting accuracy. Finally, we believe that the proposed methodology along with the results will contribute a benchmark on forthcoming malware detection proposals and research endeavors.

Our future work is directed to the following. An executable file is composed of several sections such as header, signature, and address. It might possible that some sections of that executable are irrelevant, i.e., failed to capture significant patterns to detect a malware. Therefore, identifying those relevant (irrelevant) sections will significantly reduce the analysis time, keeping the same accuracy.

Acknowledgements We would like to thank Tata Consultancy Service Innovation Lab for their Monetary support.

References

1. Nataraj, L., Karthikeyan, S., Jacob, G., Manjunath, B.S.: Malware images: visualization and automatic classification. In: Proceedings of the 8th International Symposium on Visualization for Cyber Security, p. 4. ACM (2011)
2. McGraw, G., Gregory Morrisett, J.: Attacking malicious code: a report to the infosec research council. IEEE Softw. **17**(5), 33–41 (2000)
3. Christodorescu, M., Jha, S., Seshia, S.A., Song, D.X., Bryant, R.E.: Semantics-aware malware detection. In: 2005 IEEE Symposium on Security and Privacy (S&P 2005), 8–11 May 2005, Oakland, CA, USA, pp. 32–46 (2005)
4. Vasudevan, A., Yerraballi, R.: Spike: engineering malware analysis tools using unobtrusive binary-instrumentation. In: Twenty-Nineth Australasian Computer Science Conference (ACSC2006), Computer Science 2006, Hobart, Tasmania, Australia, 16–19 Jan 2006, pp. 311–320 (2006)
5. Shannon, C.E.: A mathematical theory of communication. Bell Syst. Tech. J. **27**(Parts I and II), 379–423 623–656 (1948)
6. Shang, S., Zheng, N., Xu, J., Xu, M., Zhang, H.: Detecting malware variants via function-call graph similarity. In: 2010 5th International Conference on Malicious and Unwanted Software (MALWARE), pp. 113–120. IEEE (2010)
7. Santos, I., Brezo, F., Nieves, J., Penya, Y.K., Sanz, B., Laorden, C., Bringas, P.G.: Idea: Opcode-sequence-based malware detection. In: International Symposium on Engineering Secure Software and Systems, pp. 35–43. Springer (2010)
8. Abou-Assaleh, T., Cercone, N., Keselj, V., Sweidan, R.: N-gram-based detection of new malicious code. In: Proceedings of the 28th Annual International Computer Software and Applications Conference, 2004. COMPSAC 2004, vol. 2, pp. 41–42. IEEE (2004)
9. Tabish, S.M., Shafiq, M.Z., Farooq, M.: Malware detection using statistical analysis of byte-level file content. In: Proceedings of the ACM SIGKDD Workshop on CyberSecurity and Intelligence Informatics, Paris, France, 28 June 2009, pp. 23–31 (2009)
10. Bergeron, J., Debbabi, M., Desharnais, J., Ktari, B., Salois, M., Tawbi, N.: Detection of malicious code in cots software: a short survey. In: 1st International Software Assurance Certification Conference (ISACC99) (1999)
11. Szor, P.: The Art of Computer Virus Research and Defense. Symantec. Press (2005)
12. Oliva, A., Torralba, A.: Modeling the shape of the scene: a holistic representation of the spatial envelope. Int. J. Comput. Vis. **42**(3), 145–175 (2001)
13. Torralba, A., Murphy, K.P., Freeman, W.T., Rubin, M.A., et al.: Context-based vision system for place and object recognition. In: ICCV, vol. 3, pp. 273–280 (2003)
14. Chowdhury, G.: Introduction to Modern Information Retrieval. Facet Publishing (2010)
15. Han, K.S., Kim, S.-R., Im, E.G.: Instruction frequency-based malware classification method1. Int. Inf. Inst. (Tokyo) Inf. **15**(7), 2973 (2012)

16. Sekar, R., Bendre, M., Dhurjati, D., Bollineni, P.: A fast automaton-based method for detecting anomalous program behaviors. In: 2001 IEEE Symposium on Security and Privacy, 2001. S&P 2001. Proceedings, pp. 144–155. IEEE (2001)
17. Ye, Y., Wang, D., Li, T., Ye, D.: IMDS: intelligent malware detection system. In: Proceedings of the 13th ACM SIGKDD International Conference on Knowledge Discovery and Data Mining, pp. 1043–1047. ACM (2007)
18. Hofmeyr, S.A., Forrest, S., Somayaji, A.: Intrusion detection using sequences of system calls. J. Comput. Secur. **6**(3), 151–180 (1998)
19. Bose, A., Hu, X., Shin, K.G., Park, T.: Behavioral detection of malware on mobile handsets. In: Proceedings of the 6th International Conference on Mobile Systems, Applications, and Services, pp. 225–238. ACM (2008)
20. Wagener, G., Dulaunoy, A., et al.: Malware behaviour analysis. J. Comput. Virol. **4**(4), 279–287 (2008)
21. Bailey, M., Oberheide, J., Andersen, J., Morley Mao, Z., Jahanian, F., Nazario, J.: Automated classification and analysis of internet malware. In: International Workshop on Recent Advances in Intrusion Detection, pp. 178–197. Springer (2007)
22. Kephart, J.O.: A biologically inspired immune system for computers. In: Artificial Life IV: Proceedings of the Fourth International Workshop on the Synthesis and Simulation of Living Systems, pp. 130–139. MIT Press (1994)
23. Schultz, M.G., Eskin, E., Zadok, E., Stolfo, S.J.: Data mining methods for detection of new malicious executables. In: 2001 IEEE Symposium on Security and Privacy, Oakland, California, USA, 14–16 May 2001, pp. 38–49 (2001)
24. Kolter, J.Z., Maloof, M.A.: Learning to detect and classify malicious executables in the wild. J. Mach. Learn. Res. **6**, 2721–2744 (2006)
25. Lyda, R., Hamrock, J.: Using entropy analysis to find encrypted and packed malware. IEEE Secur. Priv. **5**(2), 40–45 (2007)
26. Li, W.-J., Wang, K., Santos, I., Herzog, B.: Fileprints: identifying filetypes by n-gram analysis. In: Information Assurance Workshop, USA, pp. 67–71. IEEE Press (2005)
27. Stolfo, S.J., Wang, K., Li, W.-J.: Towards stealthy malware detection. In: Malware Detection, pp. 231–249 (2007)
28. Stibor, T.: A study of detecting computer viruses in real-infected files in the n-gram representation with machine learning methods. In: Trends in Applied Intelligent Systems—23rd International Conference on Industrial Engineering and Other Applications of Applied Intelligent Systems, IEA/AIE 2010, Cordoba, Spain, 1–4 June 2010, Proceedings, Part I, pp. 509–519 (2010)
29. Santos, I., Nieves, J., Bringas, P.G.: Semi-supervised learning for unknown malware detection. In: International Symposium on Distributed Computing and Artificial Intelligence, pp. 415–422. Springer (2011)
30. Han, K.S., Lim, J.H., Kang, B., Im, E.G.: Malware analysis using visualized images and entropy graphs. Int. J. Inf. Secur. **14**(1), 1–14 (2015)
31. Kapur, J.N., Sahoo, P.K., Wong, A.K.C.: A new method for gray-level picture thresholding using the entropy of the histogram. Comput. Vis. Graph. Image Process. **29**(3), 273–285 (1985)
32. Campbell, F.W., Robson, J.G.: Application of Fourier analysis to the visibility of gratings. J. Physiol. **197**(3), 551 (1968)
33. Gandotra, E., Bansal, D., Sofat, S.: Malware analysis and classification: a survey. J. Inf. Secur. **5**(02), 56 (2014)
34. Microsoft Malware Classification Challenge (big 2015) (2015)

Big Data Security and Privacy: New Proposed Model of Big Data with Secured MR Layer

Priyank Jain, Manasi Gyanchandani and Nilay Khare

Abstract The publication and dispersal of crude information are urgent components in business, scholarly, and restorative applications. With an expanding number of open stages, for example, informal communities and cell phones from which information might be gathered; the volume of such information have likewise expanded after some time progressed toward becoming as Big Data. The traditional model of Big Data does not specify any level for capturing the sensitivity of data both structured and combined. It additionally needs to incorporate the notion of privacy and security where the risk of exposing personal information is probabilistically minimized. This paper introduced security and privacy layer between HDFS and MR Layer (MapReduce) known as new proposed Secured MapReduce (SMR) Layer and this model is known as SMR model. The core benefit of this work is to promote data sharing for knowledge mining. This model creates a privacy and security guarantee and data utility for data miners. In this model, running time, CPU usage, Memory usage, and Information loss are less as compared to traditional approaches.

Keywords Big Data · Security and privacy · Privacy preserving · HDFS
MR layer

List of Abbreviations

SMR Layer	Secured MapReduce Layer
MR	MapReduce
KVP	Key-Value Pairs
HDFS	Hadoop Distributed File System
API	Application Programming Interface

P. Jain (✉) · M. Gyanchandani · N. Khare
Department of Computer Science and Engineering, MANIT, Bhopal, MP, India
e-mail: Priyankjain1984@gmail.com

© Springer Nature Singapore Pte Ltd. 2019
R. Chaki et al. (eds.), *Advanced Computing and Systems for Security*,
Advances in Intelligent Systems and Computing 883,
https://doi.org/10.1007/978-981-13-3702-4_3

1 Introduction

1.1 Data Security and Privacy

Data security, according to the common definition, is the confidentiality, integrity, and availability of data. It is the practice of ensuring that the data being stored is safe from an unauthorized access and use, ensuring that the data is reliable and accurate and that it is available for use when it is needed. An information security design incorporates features, for example, gathering just the required data, protecting it, and obliterating any data that is never again required. These means will enable any business to meet the legal obligations of possessing sensitive data [1]. Privacy, on the other hand, is the appropriate use of information. In other words, merchants and companies should use the data provided to them only for the intended purpose. For example, if I make a purchase from XYZ Company and provide to them payment and address information for shipping the product, they cannot then sell my information to an unauthorized party. Organizations need to order an information security strategy for the sole purpose of guaranteeing information protection or the protection of their consumer's data. All the more in this way, organizations must guarantee information security in light of the fact that the data is an advantage for the organization. A data security policy is essentially the way to the coveted end, which is information protection. Nonetheless, no information security strategy can defeat the ready offer or request of the buyer information that was depended to an organization [1, 2].

1.2 The Need for Privacy and Security in Big Data

The term Big Data alludes to the massive information of advanced data organizations and government gather about us and our environment. Consistently, worldwide make quintillion bytes of information so much that 90% of the information on the world today has been made over the most recent 2 years alone. Security and privacy issues are magnified by velocity, volume, and variety of Big Data, for instance, immense scale cloud foundations, diversity of data sources and formats, streaming nature of information procurement, and high volume bury cloud relocation. The utilization of huge scale uproarious foundation, with differences of programming stages, spread crosswise over expansive systems of PCs, additionally, increment the assault arrangement of the whole framework. The main security and privacy challenges in Big Data [3–5]:

a. Secure computations in distributed program frameworks.
b. Security best practices for non-relational data sources.
c. Secure data sources and transition logs.

d. End-point input validation/filtering.
e. Real-time security/compliance monitoring.
f. Versatile and composable privacy-data mining and analytics.
g. Cryptographically enforced access control and secure communication.
h. Granular access control.
i. Granular audits.
j. Data provenance.

1.3   The Need for Light Weight Encryption

Heavy and lightweight encryption algorithms are used for secured communication over the Internet. Lightweight encryption algorithms are preferred over heavyweight encryption algorithms in low power designs and devices mainly because of their reduced resource requirements such as memory and execution time. A lightweight encryption technique takes less time for encryption and provides better security than existing heavyweight algorithms such as AES, RSA, PGP, TEA, and RC6 [6]. The proposed solution employs multi-level lightweight encryption along with key encryption and thereby decreases the possibility of various threats by attackers. In recent years, a large amount of person-specific data was collected by both government and private entities. Laws and regulations require that some collected data must be made public (Example: Census and Healthcare Data).

1.4   Proposed Secure MapReduce (SMR) Model

The traditional model of Big Data [7–11] does not specify any level for capturing the sensitivity of data both structured and combined. It additionally needs to consolidate the thought of protection and security where the danger of uncovering individual data is probabilistically limited. Given the high volume of enormous information, and the combination of structured and unstructured data requires some set of new models for Big Data so as to increase privacy and security. These algorithms build on current privacy-preserving data techniques, thus come up with a new model which incorporates a new layer of privacy on the MapReduce phase of Big Data architecture. This new layer thus implemented the security algorithms on the data individually as the data come across the MapReduce phases. The security algorithm should be lightly weighted encryption techniques so that the overhead of new algorithms do not affect the main functioning of Big Data. The data thus can be protected and secured, when it is processed through this new proposed Secured MapReduce (SMR) Layer of Big Data. It starts from a collection of data from weblogs, social data,

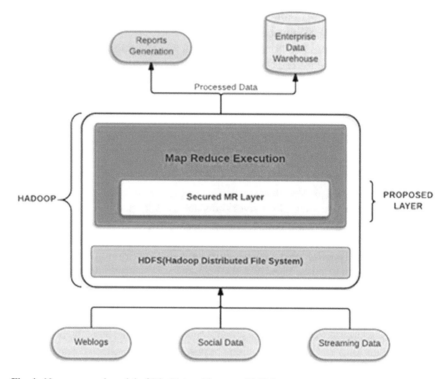

Fig. 1 New proposed model of Big Data with secured MR layer

streaming data, and then the collected data is sent to HDFS (Hadoop Distributed File System) [12–16]. This proposed model introduced privacy layer between HDFS and MR Layer (MapReduce) known as Secured MapReduce (SMR) Layer in Fig. 1. To increase security and privacy of the data, perturbation and randomized techniques were used.

The succeeding sections are: Sect. 2 is the related work, Sect. 3 describes proposed work using SMR, Sect. 4 shows data set description, Sect. 5 represents implementation and result, and Sect. 6 concludes the presented work.

2   Related Work

Evfimievski [17] work refers to randomized based privacy approach for a small amount of data. In this work, client having personal information and the server is interested in only collecting, statistically important, properties of this information. The clients can guard the privacy of their data by unsettling the data with a randomization algorithm and further giving in the randomized version. It describes certain ways and consequences in randomization for numerical and categorical data and argues the concern of measuring privacy. The exploration in utilizing randomization for preserving privacy gives an impression of being a piece of some more profound measurable way to deal with security and privacy, providing a connection to the groundbreaking work of Claude Shannon on secrecy systems, and enables us to look at privacy under a different angle than the conventional cryptographic approach. Randomization does not count on complexity hypotheses from algebra or number theory and does not need costly cryptographic operations or sophisticated protocols. It is likely that future studies will combine statistical approach to privacy with cryptography and secure multiparty computation, to the mutual benefit of all of them. Roy et al. [18] work refer to a MapReduce-based system that provides strong privacy and security and provides assurances for distributed computations on sensitive data known as Airavat model. This model comprises of a novel integration of mandatory access control (MAC) and differential privacy (DP). Here, data providers control the security plan for their sensitive data, including a mathematical assured on potential privacy violations. Airavat considered the first model which integrates MAC with DP and enables many privacy-preserving MapReduce calculations without the need of audit of un-trusted code. Tripathy and Mitra [19] work presents a procedure which can be used to achieve k-anonymity and l-diversity in social network anonymization. This algorithm is grounded in some past algorithms developed in this direction. This algorithm is altered reasonably from their corresponding algorithm for microdata and also relies upon some modified algorithms developed for anonymization against neighborhood attack. The algorithm still needs few advancements in order to decrease the complexity in order that it can be applied to large social networks. The p-sensitivity issue as specified by Machanavajjhala is yet to be dealt with so far even in the relational database case. Only the distinct A diversity has been measured and utilized up to this point. Mohammadian et al. [20] work describe Fast Anonymization of Big Data Streams (FAST) algorithm is an anonymization algorithm which is used to boost up anonymization of a Big Data streams parallel algorithm which introduces an efficient Big Data anonymization by a multi-thread technique to shield the data against attacks on privacy. It results in the efficiency of anonymizing Big Data and also decreases the degree of information loss and cost metric. According to the Terzi et al. [21] work, network traffic should be expressed in code with suitable standards, access to devices should be checked, employees should be authorized to access systems, analysis should be done on anonymized data sending and receiving should be made for the secure channel to prevent data drip, and network should be observed for threats. As per Derbeko et al. [22] work when a MapReduce computation is implemented in public or hybrid clouds, privacy, a safety of data, and output of MapReduce are essentially considered. In public and hybrid clouds environment, implementation of MapReduce paradigm requires privacy, integrity, and correctness of the outputs as well as verification of mappers reducers. Mehmood et al. [2] work provides a complete outline of the privacy preservation techniques in Big Data and presents the challenges of existing mechanisms. They explained the infrastructure of Big Data and the privacy maintaining techniques in each phase of the Big Data lifecycle. A few future study aspects for Big Data privacy are discussed as under access control, safe end-to-end communication, data anonymization, decentralized storage effective machine learning techniques and distributed data analytics secure combined computation techniques such as homomorphic encryption can be deployed to solve such issues. The fundamental difficulty in deploying homomorphic encryption in the framework of Big Data analytics is to keep the computational complexity as low as possible. Kacha and Zitouni [23] work describes principal issues related to data security that is raised by cloud environment are classified into three categories: (1) data security issues raised by single cloud characteristics, (2) data security issues raised by data lifecycle in cloud computing, (3) data security issues associated to data security attributes (confidentiality, integrity, and availability). As per Ilavarasi [24] work, privacy has become an important concern while publishing microdata about a population. The emerging area called privacy-preserving data mining (PPDM) concentrates on any person-specific privacy without negotiating data mining results. The enhancing growth of PPDM algorithms enhances the concept of investigating the privacy inferences and the cross-cutting issues between privacy versus the utility of data.

3 Proposed Work

3.1 Overview

The enterprises are facing deployment and management challenges with Big Data. Hadoop's center determination is as yet being created by the Apache people group [25] and, up to this point, do not palatably address endeavor necessities for powerful security, arrangement implementation, and administrative consistency. While Hadoop may have its challenges, its approach, which allows for the distributed processing of large data sets across clusters of computers, represents the future of enterprise computing. With a specific end goal to fill the security holes that exist in all open-source Hadoop dispersions, a strong pathway for securing circulated processing situations in the endeavor is given by utilizing randomization strategy for the security keeping up the integrity of the specifications [26–30].

3.2 Concept of Randomization

Randomization [31] is a well-known technique for masking the data utilizing random noise while making sure that the random noise still preserves the signal from data so that the patterns can still be precisely assessed. The noise added is sufficiently large so that individual record values cannot be recouped. Therefore, techniques are designed to derive aggregate distributions from the perturbed records. Statistics, randomized calculation, and many other related fields are full of theorems, laws, and algorithms that rely on probabilistic characterization of random processes that often work quite accurately.

Original values $x_1; x_2 \dots x_n$ from probability distribution function X. To hide these values, we use $y_1; y_2 \dots y_n$ from probability distribution Y. Given $x_1 + y_1; x_2 + y_2; \dots; x_n + y_n$, the probability distribution of Y estimates the probability distribution of X.

Uniform distribution: The arbitrary variable has a uniform appropriation over an interim $[-\alpha, +\alpha]$. The mean of the arbitrary variable is 0.

$F_x^0 :=$ Uniform distribution $j := 0$ //Iteration number.

The Bayes Reconstruction Method used for reconstruction purpose is shown in Eq. 1.

$$f_{x^{j+1}}(a) = \frac{1}{n} \sum_{i=1}^{n} \frac{f_y((x_i + y_i) - a) f_x^j(a)}{\int_{-\infty}^{\infty} f_y((x_i + y_i) - a) f_x^j(a)} \tag{1}$$

$J := j + 1$
Until (stopping criterion met).

3.3 Perturbing the Data

The class of techniques for privacy-preserving data mining by randomly perturbing the data while preserving the underlying probabilistic properties. It investigates the arbitrary esteem perturbation-based approach, an outstanding system for covering the information utilizing irregular noise. This approach tries to save information security by including arbitrary commotion while ensuring that the random noise still preserves the original data from the data so that the patterns can still be precisely evaluated. People doubt the utility of the arbitrary esteem contortion strategy in protection conservation, and they take note of that irregular items (especially arbitrary grids) have 'predictable' structures in the spectral area, and it builds up an irregular lattice-based phantom sifting strategy to recover unique information from the dataset contorted by including arbitrary esteems. This predominantly manages the privacy of the information. There are two conceivable methodologies for perturbing the information. They are value-class membership and value distortion [1, 32–35].

3.4 Hash Mapping

HashMap keeps up key and esteem sets and regularly indicated as HashMap <Key;Value> or HashMap <K;V>. HashMap executes map interface. HashMap is like hash table with two special cases hash map techniques are unsynchronized and it permits invalid key and invalid esteems dissimilar to hash table. It is utilized for keeping up key and esteem mapping. It is not an ordered collection which means it does not return the keys and values in the same order in which they have been inserted into the hash map. It neither does any type of sorting to the stored keys and values.

a. **Java HashMap class**

A HashMap contains values in view of the key. It actualizes the map interface and expands AbstractMap class. It contains just extraordinary components. It might have one invalid key and different invalid esteems. It keeps up no request.

3.5 Encryption

In this process, original data is passed to HDFS (Hadoop Distributed File System) then the data from HDFS will be passed to MapReduce layer shown in Fig. 2. The original data may be in the form of weblogs, streaming data, social data. The encryption process takes place in MapReduce layer. Once the data entered the map reduce layer, the encryption starts. Encryption is the step where encode or encrypt the given data. There are two levels involved in this encryption. In first level converting text data to number, to do this first consider text and dividing each word text into tokens. It will take key-value pairs (KVP) model, by considering each unique word and counting the number of times the word is repeating in the given data, where the key is representing each unique word and value is representing a number of times the word is repeating. This process not only provides lightweight encryption but also provide high privacy of given data. The second level performs randomization process in converted number data, which enhance the privacy level. HybrEx model [36] shown in Fig. 3 has been implemented where the data is first processed in the private cloud at the time of encryption and then the data is processed in public cloud at the time of decryption.

3.6 Decryption

Decryption shown in Fig. 4 is the step where decrypt the encrypted data. It is the reverse process of encryption. In this process, processed data (key-value pairs) is passed to HDFS then the output of this is passed into MapReduce layer. Where

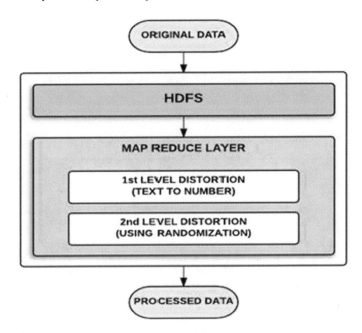

Fig. 2 New proposed model of Big Data with secured MR layer with encryption

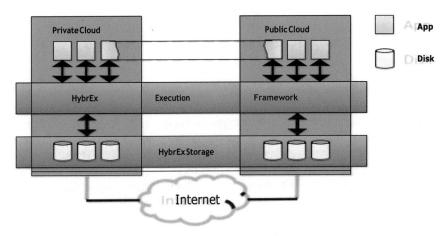

Fig. 3 Architecture of the HybrEx model

decryption takes place in MapReduce layer. This phase is called reconstruction phase. There are two levels involved in this decryption process.

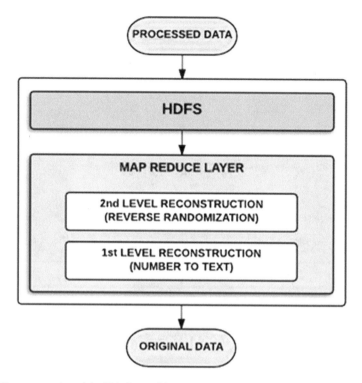

Fig. 4 New proposed model of Big Data with secured MR layer with decryption

3.6.1 Reverse Randomization

This level will use randomization to decrypt the encrypted message to some extent. Where randomization method is a privacy-preserving technique in which noise is added to the data so that the original values of records can be masked. The noise added is sufficiently large so that individual record values cannot be recovered. Therefore, techniques are designed to derive aggregate distributions from the perturbed records.

3.6.2 Converting Number to Text Data

To convert a number to text this model used a concept of Key-Value pair after the reverse randomization process. Every word (token) is key and a number of times it is repeating is value. But as order is not maintained here to maintain the order will retrieve the data from the file where wrote the order. By this, it will successfully complete the number to text data conversion. After following these steps will get the original data back.

Algorithm 1 Secure MapReduce (SMR) Encryption Algorithm

Input: File data F.

Output: Encrypted file SMR, an Encrypted file of frequencies EC.

Mapper Phase:

    1. Partition(F).

    2. for each line L_i

        A.   read();

        B.   tokenize();

        C.   for each word w_j

            i. Convert w_j to number n_j ; // here we covert the word into the number and store this into a hash map and the SMR file.

            ii. rand(n_j); // here the number is replaced by the random

    3. A. Write(rand(n_j), SMR); // this original-random number pair is written into the SMR encrypted file for the process of reverse randomization.

        B.   Write(mapper id, SMR);

Reducer Phase:

    1.   Combine(mapper result);

    2.   Count(w_j);

    3.   Encrypt(Count(w_j));

3.7 Description of SMR Encryption Algorithm

In Algorithm 1 Partition (F), the HDFS will partition the file data (F) into n number of blocks each of size 128 MB and distribute these to m nodes where mapper and reducer will work. Now by read() function, the mapper will read the part of the data file line by line, then tokenize the string into separate words using tokenize(). Then, each word will be converted to a number for one-way privacy purpose, which will be further changed to another random number through the process of randomization for which we use the function rand(), for two-way privacy purpose. The write() function would lead to this random number being written to a common file which is called the SMR encrypted file, where the pairs of random number corresponding to the original number are written and also the order of the original numbered data is written here and this file is used at the time of reverse randomization. Thus, the original numbered data is retrieved and also the order of the sentence is maintained. Here each mapper will maintain the count of each number. Now the results of all the mappers will be passed to the reducer. The file which is generated on the mapper side is the encrypted SMR file which maintains the (noisy number)−(original number) pairs and the entire order of the original numbered file along with the mapper ids. Note: the mapper task will also write the mapper id at the end of each sentence to maintain the regular order of sentences. In the reducer phase, each reducer will combine the result of mapper and maintain the count of words in the whole file. Then the frequency of each word is also encrypted here. So, there is a word which is encrypted and its frequency which is also encrypted.

3.8 Application of Query on the Encrypted Data

Analyzing the Twitter dataset could help one improve in many spheres like marketing companies for increases the popularity of their products, by many surveys for seeing who is most influential currently, for finding out latest trends and patterns which in turn would help to enhance profit and business. For applying any of the above queries, the component used in this paper is the hive. After the data is transferred from HDFS to MapReduce, where it is encrypted, the component hive comes into play and answers these queries. In this way, the identity of the person is preserved along with the queries being answered. Storing the data into HDFS and then processing the data through a secured MapReduce phase where encryption process is performed, all this is done in a private cloud and then the encrypted data is given to the public cloud where only that person can get the data who have the decryption key. The decryption key is nothing but the SMR encrypted file which contains (noisy number)−(original number) pairs and the order of the entire original numbered file and also the mapper ids along with the hash map. Now if someone tries to query the data using hive, the person will get an encrypted answer and hence in this way this work provided two-way security to the data. Only that person having this decryption key can get the information.

Algorithm 2 Secure MapReduce (SMR) Decryption Algorithm

Input: Encrypted file of words SMR, Encrypted file of frequencies EC.
Output: Decrypted file of words D, a Decrypted file of frequencies DC.
    Mapper Phase:

    1. Partition(SMR);.

    2. for each line L_i

        A. read(mapper id);

        B. add mapper id to hash map(H);

        C. tokenize();

        D. S = reverse randomization(number); // reverse randomization is done
           with the help of the SMR encrypted file.

        E. add string S to hash map(H);

    3. Decrypt(C);

    Reducer Phase:

        1. read(hash map(H));

        2. generate(D);

        3. generate(DC);

3.9   Description of SMR Decryption Algorithm

In Algorithm 2 firstly, the server will receive the encrypted file from the client through a network connection. Now this encrypted file, SMR, will act as input to the server cluster. Now HDFS at this side will first partition the file data into l blocks and distribute them to several nodes again (Partition (SMR)). The partition of a file will be again read line by line. It will read the mapper id first and create a hash map based on these mapper ids. The mapper will read one line and again tokenize it and then decrypt the number into the corresponding word by the process of reverse randomization. And further, add the whole decrypted string to the hash map under the matching mapper id. Likewise, all the mappers will add their decrypted strings to the same hash map. And now this hash map will be passed onto the reducer side. Also, the file containing the word and its frequency will also be decrypted and pass onto the reducer end. The reducer will again perform two tasks simultaneously. It will first read the hash map line by line and generate a decrypted file which will contain the whole data in order (D) and it will also combine the results of mapper to generate an output file of words and their frequency (DC).

4 Dataset Description

4.1 Accessing the Twitter API

The way that scientists and other individuals who need to get huge publicly accessible Twitter datasets [37] are through their API. Programming interface remains for application programming interface and many administrations that need to begin an engineering group around their item generally discharges one. Facebook has an API that is to some degree prohibitive, while Klout has an API to let you naturally look into Klout scores and all their distinctive facets. The Twitter API has two unique types: RESTful and Streaming. The RESTful API is helpful for getting things like arrangements of supporters and the individuals who take after a specific client and is the thing that most Twitter customers are worked off of. We will concentrate on the Streaming API. The Streaming API works by making a demand for a particular kind of information sifted by watchword, client, geographic range, or an irregular specimen and afterward keeping the association opens the length of there are no blunders in the association. For this purposes, using the tweepy bundle to get to the Streaming API.

4.2 Understanding Twitter Data

When associated with the Twitter API, select RESTful API or Streaming API, to begin recovering a bundle of information. The information gets back will be encoded in JSON or JavaScript Object Notation. JSON is an approach to encode convoluted data in a stage autonomous manner. It could be viewed as the most widely used language of data trade on the Internet. At the point when clicking a sweet Web 2.0 catch on Facebook or Amazon and the page delivers a lightbox (a container that floats over a page without leaving the page you're on now), there was presumably some JSON included. JSON is a fairly shortsighted and rich approach to encode complex information structures. At the point when a tweet returns from the API, this is the thing that it would appear that. So that to simply remove helpful data from it promptly.

4.3 Collecting Data

The initial step is to get a duplicate of tweepy [37] (either by looking at the store or simply downloading it) and introducing it. The following steps to do are, firstly it has to make an occasion of a tweepy StreamListener to deal with the approaching information. Begin another document for every 20,000 tweets, labeled with a prefix and a timestamp. This record is 'slistener.py'. Next, need the script that does the gathering itself. This record is 'streaming.py', can gather on clients, watchwords, or particular areas characterized by bouncing boxes. The API documentation has more data on this. For the present, some well-known keywords like Delhi and India, etc. are being used in the proposed model (keywords are case-insensitive).

5 Implementation and Results

The platform used for the deployment is HP Z840 workstation. It is having 64-bit dual-core processors and 8 GB of RAM. Each workstation has 40 Cores. Forty Cores are used for name node and 160 cores are used for data nodes for implementing SMR.

5.1 Encryption Side Implementation

When the input file is provided to the master node, here HybrEx model [36] has been implemented where the data is first processed in the private cloud at the time of Encryption. Hadoop mechanism [25, 38–41] first partition the whole file into small parts shown in Fig. 5. Then these parts are distributed to many mapper tasks by the job tracker. Inside each mapper, every word of a sentence gets encrypted with the specified logic of randomization and is written to a file along with the mapper id of the corresponding mapper task of the slave node. Now the reducer task will aggregate the results of all mapper tasks and finally generate the encrypted file as output which is to be transmitted to the public cloud.

5.2 Decryption Side Implementation

Anyone accessing this data from the public cloud will always get an encrypted answer when the person queries the encrypted data. Only that individual who is having the decryption key will be able to get the original data and the desired outputs to the queries. Now, what happens when the person tries to decrypt the data with the decryption key? Now, when the encrypted file is provided to the master node, the Hadoop mechanism partitions the whole file into small parts shown in Fig. 6.

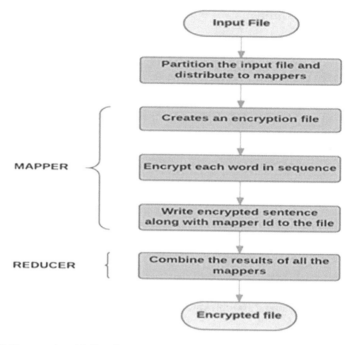

Fig. 5 SMR encryption side flowchart

Then these parts are distributed to many mapper tasks by the job tracker. Inside each mapper, every word of a sentence gets decrypted with the specified logic of randomization and a hash map is created according to the mapper id and all the sentences belonging to the one mapper are placed at one place. Now the reducer task will finally write the whole hash map in order to a file and generate the decrypted file as output which is same as the original file.

5.3 Results

5.3.1 Time Taken with Respect to Data Size (With SMR Layer, Without SMR Layer)

In Fig. 7 graph shows that when security is implemented through Hadoop SMR layer, the range of data starts from 1 to 5 GB. SMR takes bit more time as compared to the approaches without SMR layer. SMR running time difference decrease when increasing the data size. To execute 2 GB of data without Secured MR layer take 47.2 ms as compared with Secured MR layer approach take 70.9 ms to execute same data. This time difference decreases to execute 5 GB of data, without SMR layer take 76 ms as compared with Secured MR layer approach take only 85 ms to execute

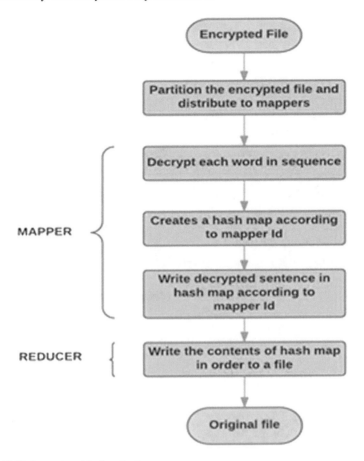

Fig. 6 SMR decryption side flowchart

same data. SMR layer approach is suitable for Big Data when increase the data size time difference get minimized.

5.3.2 Time Taken with Respect to a Number of Cores

In Fig. 8 graph shows that the parallel execution of Hadoop, SMR layer tasks markedly reduces the time when it comes to increasing the number of cores 40–160 as the nodes will simultaneously perform with the same speed of processing the data. Using 40 Cores to execute 5 GB of data takes 210 ms and using 160 Cores to execute the same amount of data take only 28 ms.

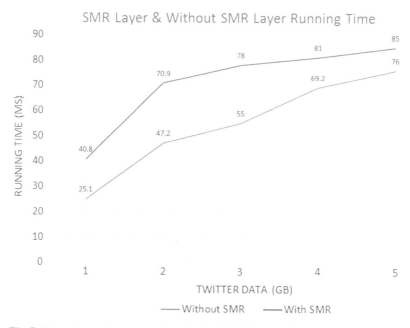

Fig. 7 Time taken with respect to data size (with SMR layer, without SMR layer)

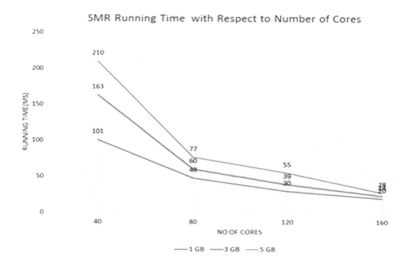

Fig. 8 Time taken with respect to number of cores

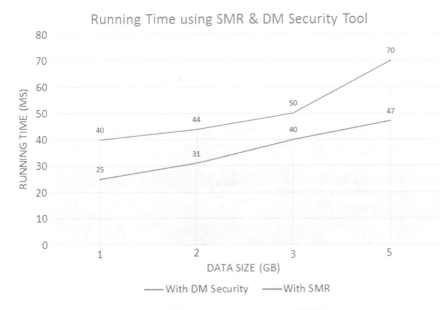

Fig. 9 Time taken with respect to DM encryption versus secured MR layer

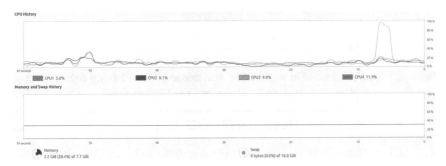

Fig. 10 CPU and memory uses in SMR

5.3.3 Time Taken with Respect to SMR Layer versus Traditional Data Mining Encryption Algorithm Implemented Security

In Fig. 9, graph shows that the traditional methods of data mining encryption algorithm take much more time when compared with proposed Hadoop SMR layer in implementing the security algorithms. Also as the size of the data increases, Hadoop method seems to take less time than traditional data mining encryption methods [23]. Table 1 represents the comparative study of results with respect to running time.

Table 1 Comparative study of results with respect to running time

Data size	Time taken with respect to data size (ms)		Time taken with respect to No. of cores (ms)			
Data volume in GB	With SMR	Without SMR	40 Cores	80 Cores	120 Cores	160 Cores
1	40.8	25.1	101	48	30	20
3	78	55	163	60	39	24
5	85	76	210	77	55	28

Fig. 11 Information loss with respect to traditional method versus secured MR layer

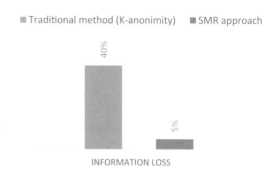

5.3.4 CPU and Memory Uses in SMR

SMR take less CPU uses in multi-node environment, here having four HP Z840 workstation 64-bit dual-core processors, the range of CPU uses between 5 and 11.9% shown in Fig. 11. Memory uses during the running of SMR is 28.4%, i.e., 2.2 GB of 7.7 GB and swap is 0%. The result shows SMR layer approach is suitable for Big Data.

5.3.5 Information Loss

A release of data is said to have the k-anonymity property if the information for each person contained in the release cannot be distinguished from at least k − 1 individuals whose information also appear in the release. The information loss in a traditional method such as K-anonymity [42, 43] is around 40% and that in this new proposed approach SMR layer is reduced to only 5% as shown in Fig. 10. Therefore, this new SMR layer is an improvement over the existing methods.

6 Conclusion and Future Work

This paper discussed the Big Data aspects with a focus on privacy and security. This paper proposes a methodology to protect Big Data information by the proposed SMR model, and it likewise needs to consolidate the thought of protection and security where the danger of uncovering individual data is probabilistically limited. SMR model is based on lightweight encryption using randomization and perturbation method of security maintaining the integrity of the specifications. Traditional methods of data mining encryption algorithm take much more time when compared to the proposed SMR model in implementing the security algorithms. The result shows this approach is an advantage for Big Data, with better privacy and security with a minute difference in additional processing time. When increasing the data size, processing time difference gets minimized as compared to traditional approaches (without SMR layer). Experiment results show that CPU usage, memory usage, and information loss are less in the proposed SMR layer. Future work will focus on the security of real-time Big Data information.

Acknowledgements We acknowledge the support of Madhya Pradesh Council of India. We are also thankful to Dr. Rajesh Wadhvani and Dr.Sri Khetwat Saritha for providing high configuration system facilities form their respective laboratory Information retrieval lab and Machine Learning lab of MANIT Bhopal.

References

1. Jain, P., Gyanchandani, M., Khare, N.: Big data privacy: a technological perspective and review. J. Big Data **3** (2016). ISSN 2196-1115
2. Mehmood, A., Natgunanathan, I., Xiang, Y., Hua, G., Guo, S.: Protection of big data privacy. IEEE Access **4**, 1821–1834 (2016). https://doi.org/10.1109/access.2016.2558446
3. Big Data Top Challenge 2016. https://downloads.cloudsecurityalliance.org/initiatives/bdwg/BigDataTopTenv1.pdf
4. Big Data Submits. https://theinnovationenterprise.com/summits/big-data-innovation-mumbai/eventactivities=5546
5. The intersection of privacy and security data privacy day event 2012. https://concurringopinions.com/archives/2012/01/the-intersection-of-privacy-and-security-data-privacy-day-event-at-gw-law-school.html
6. Savas, O., Deng, J.: Book Titled Big Data Analytics in Cybersecurity. CRC Press, Taylor Francis Group
7. Sagiroglu, S., Sinanc, D.: Big data: a review. J. Big Data 20–24 (2013)
8. Chavan, V., Phursule, R.N.: Survey paper on big data. Int. J. Comput. Sci. Inf. Technol. **5**(6) (2014)
9. Groves, P., Kayyali, B., Knott, D., Kuiken, S.V.: The Big Data Revolution in Healthcare. McKinsey & Company, New York (2013)
10. Lin, J.: MapReduce is good enough the control project. IEEE Comput. **32** (2013)
11. Patel, A.B., Birla, M., Nair, U.: Addressing big data problem using Hadoop and Map Reduce. In: Nirma University International Conference on Engineering in Proceedings (2012)
12. Acampora, G., et al.: Data analytics for pervasive health. In: Healthcare Data Analytics (2015). ISSN 533-576

13. Kulkarni, A.P., Khandelwal, M.: Survey on Hadoop and introduction to YARN. Int. J. Emerg. Technol. Adv. Eng. **4**(5) (2014). www.ijetae.com. ISSN 2250-2459
14. Yu, E., Deng, S.: Understanding software ecosystems: a strategic modeling approach. In: Proceedings of the Workshop on Software Ecosystems 2011, IWSECO-2011, pp. 6–16 (2011)
15. Shim, K: MapReduce algorithms for big data analysis. In: DNIS. LNCS, pp. 44–48 (2013)
16. Arora, S., Goel, M.: Survey paper on scheduling in Hadoop. Int. J. Adv. Res. Comput. Sci. Softw. Eng. **4**(5) (2014)
17. Evfimievski, S.: Randomization techniques for privacy preserving association rule mining. In: SIGKDD Exploration, vol. 4, no. 2 (2002)
18. Roy, I., Ramadan, H.E., Setty, T.V., Kilzer, A., Shmatikov, V., Witchel, E.: Airavat: security and privacy for MapReduce. In: Castro, M. (eds.) Proceedings of the 7th USENIX Symposium on Networked Systems Design and Implementation, pp. 297–312. USENIX Association, San Jose (2010)
19. Tripathy, K., Mitra, A.: An algorithm to achieve k-anonymity and l-diversity anonymization in social networks. In: Proceedings of Fourth International Conference on Computational Aspects of Social Networks (CA-SoN), Sao Carlos, pp. 126–131 (2012)
20. Mohammadian, E., Noferesti, M., Jalili, R.: FAST: fast anonymization of big data streams. In: Proceedings of the 2014 International Conference on Big Data Science and Computing, p. 23 (2014)
21. Terzi, D.S., Terzi, R., Sagiroglu, S.: A survey on security and privacy issues in big data. In: Proceedings of ICITST 2015, London, UK, Dec 2015
22. Derbeko, P., et al.: Security and privacy aspects in MapReduce on clouds: a survey. Comput. Sci. Rev. **20**, 1932–128 (2016)
23. Kacha, L., Zitouni, A.: An overview on data security in cloud computing. In: CoMeSySo: Cybernetics Approaches in Intelligent Systems, pp. 250–261. Springer (2017)
24. Ilavarasi, K., Sathiyabhama, B.: An evolutionary feature set decomposition based anonymization for classification workloads: privacy preserving data mining. J. Cluster Comput. (2017)
25. ApacheHDFS. http://hadoop.apache.org/hdfs
26. Sweeney, L.: K-anonymity: a model for protecting privacy. Int J Uncertain Fuzz. **10**(5), 55770 (2002)
27. Zakerdah, H., Aggarwal, C.C., Barker, K.: Privacy-Preserving Big Data Publishing. ACM, La Jolla (2015)
28. Morey, T., Forbath, T., Schoop, A.: Customer data: designing for transparency and trust. Harv. Bus. Rev. 93–95 (2015)
29. Friedman, A., Wolff, R., Schuster, A.: Providing k-anonymity in data mining. Int. J. Very Large Data Bases **17**(4), 789–804 (2008)
30. Fung, B., et al.: Privacy-preserving data publishing: a survey of recent developments. ACM Comput. Surv. (CSUR) 42–44 (2010)
31. Cevher, V., Becker, S., Schmidt, M.: Convex optimization for big data: scalable, randomized, and parallel algorithms for big data analytics. IEEE Signal Process. Mag. **31**(5), 32–43 (2014)
32. Kuo, M.H., Sahama, T., Kushniruk, A.W., Borycki, E.M., Grunwell, D.K.: Health big data analytics: current perspectives, challenges and potential solutions. Int. J. Big Data Intell. **1**(1/2), 114–126 (2014)
33. Fung, B.C.M., Wang, K., Chen, R., Yu, P.S.: Privacy-preserving data publishing: a survey of recent developments. ACM Comput. Surv. **42**(4) (2010)
34. Jain, P., Pathak, N., Tapashetti, P., Umesh, A.S.: Privacy-preserving processing of data decision tree based on sample selection and singular value decomposition. In: 2013 9th International Conference on Information Assurance and Security (IAS), Gammarth, pp. 91–95 (2013)
35. Jain, P., Gyanchandani, M., Khare, N.: Privacy and security concerns in healthcare big data: an innovative prescriptive. J. Inf. Assur. Secur. **12**(1), 18–30 (2017)
36. Ko, S.Y, Jeon, K., Morales, R.: The HybrEx model for confidentiality and privacy in cloud computing. In: 3rd USENIX Workshop on Hot Topics in Cloud Computing, HotCloud'11, Portland (2011)

37. Tweepy Dataset Online. https://marcobonzanini.com/2015/03/02/mining-twitter-data-with-python-part-1/
38. First Things First—Highmark Makes Healthcare-Fraud Prevention Top Priority with SAS (2006)
39. Apache Hive. http://hive.apache.org
40. Borthakur, D., Sarma, J.S., Gray, J., Muthukkaruppan, K., Spiegelberg, N., Kuang, H., Krangana Than, D.M.S., Menon, A., Rash, S., Schmidt, R., Amitanand, A.: Apache Hadoop Goes Realtime at Facebook ACM SIGMOD, Athens, Greece (2011). 978-1-4503-0661-4/11/06
41. Mrigank, M., Akashdeep, K., Snehasish, D., Kumar, N.: Analysis of Big Data Using Apache Hadoop and Map Reduce, vol. 4, no. 5 (2014)
42. Ghinita, G., Karras, P., Kalnis, P., Mamoulis, N.: Fast data anonymization with low information loss. In: Proceedings of International Conference on Very Large Data Bases (VLDB), pp. 758–769 (2007)
43. Yin, C., Zhang, S., Xi, J., Wang, J.: An improved anonymity model for big data security based on clustering algorithm. In: Combined Special Issues on Security and privacy in social networks (NSS2015) and 18th IEEE International Conference on Computational Science and Engineering (CSE2015), vol. 29, Issue 7, 10 Apr 2017

Part II
IoT and Networks

Ambient Monitoring in Smart Home for Independent Living

R. Kavitha and Sumitra Binu

Abstract Ambient monitoring is a much discussed area in the domain of smart home research. Ambient monitoring system supports and encourages the elders to live independently. In this paper, we deliberate upon the framework of an ambient monitoring system for elders. The necessity of the smart home system for elders, the role of activity recognition in a smart home system and influence of the segmentation method in activity recognition are discussed. In this work, a new segmentation method called area-based segmentation using optimal change point detection is proposed. This segmentation method is implemented and results are analysed by using real sensor data which is collected from smart home test bed. Set of features are extracted from the segmented data, and the activities are classified using Naive Bayes, kNN and SVM classifiers. This research work gives an insight to the researchers into the application of activity recognition in smart homes.

Keywords Activity recognition · Smart home
Sensor · Machine learning classifiers · Segmentation

1 Introduction

In India, the joint family system is the basic social structure. Members of the family join together to take care of elders in their old age. Nowadays, the joint family system is disintegrating and nuclear families are becoming more common, because of the increase in migration of younger generation from rural areas to urban areas or to a foreign country [1]. This leads to the elders living alone. The statistical survey by United Nations Department of Economic and Social Affairs/Population Division [2] says that in India 9.5% of the population comprises of elders above 60 years. This

R. Kavitha (✉) · S. Binu
Christ University, Bangalore, Karnataka, India
e-mail: kavitha.r@christuniversity.in

S. Binu
e-mail: sumitra.binu@christuniversity.in

© Springer Nature Singapore Pte Ltd. 2019
R. Chaki et al. (eds.), *Advanced Computing and Systems for Security*,
Advances in Intelligent Systems and Computing 883,
https://doi.org/10.1007/978-981-13-3702-4_4

may reach 22.2% in 2050 and 44.4% in 2100. The population of elders are gradually increasing and on the other side, they are becoming alone in their older age due to migration of their family members. It has now become a new challenge to provide the elders safe, comfort and secure living environment. These challenges are being addressed by the smart home system. A smart home (SH) is a home where different types of smart devices or sensors are deployed to sense the daily activity of residents of the smart home. In a smart home system, the physical components sense the environment and pass the sensed information to the home intelligent system through home network and subnetworks. The home control system takes the decision and passes the control information to the actuators through a home network. For example, a gas sensor detects the gas leakage in a smart home and passes this message to the home control system through ZigBee or wireless network [3]. Control system decides to switch off the gas valve and passes this to the actuator, which will close the gas valve. Smart home system firmly provides relaxed, safe-sheltered and convenient life to the elders similar to that of an assisted living environment. Activity recognition plays a vital role in smart home system. This helps to find whether a disabled person, living in a smart home is capable of moving around [4]. Research says that the person is normal when they do their usual activities regularly without any support.

The architecture of a conventional home consists of many areas like kitchen, bedroom and bathroom. There are two ways to construct the smart home. One is making home architecture with only smart appliances which is having the capability of sensing the environment smartly and connect to other device or sources on its own. This setup fully depends on smart architecture. Other one is making SH without disturbing the available architecture and normal appliances. In this research, second type of SH is used. This idea is possible by deploying different sensors like motion sensor, temperature sensor and door sensor on the ceiling or walls of the home. Sensors deployed in a smart home sense the movement and generate the sensor stream. This sensor stream is used to understand the environment. The segmentation is applied on this sensor stream to recognize the activity happening in a SH.

2 Literature Review

The human activity recognition is the most promising research area in smart home ambient-assisted living. Activity recognition is a complex process and helps to know whether the disabled person living in a SH is capable of doing daily life tasks regularly. So, the accurate activity recognition helps to make the SH system more significant. To improve the accuracy of an activity recognition, segmentation approaches are introduced. Researchers proposed many segmentation methods. The segmentation methods have a great potential impact on increasing the success of an accurate activity recognition. As per literature review, the most popular segmentation methods are activity-based segmentation [5] and time-based segmentation [6].

Different segmentation approaches are followed by the researchers to improvise activity recognition. Broadly, there are two different approaches to recognize an activity. First one is offline mode which recognizes the activities after completion of entire data collection. The second approach, viz. online mode helps to recognize activities when they happen. Activity-based segmentation is a good example for offline-mode segmentation. In this approach, sensor streaming data are divided into segments at the point where the changes in activity are found [7]. Each and every segment probably directly relates to an activity. Many researchers approached this segmentation technique. Since it is an activity-defined approach, initial and end points are determined for each activity, prior to explicitly identifying specific activities. Time-based segmentation is a good example for online segmentation. In this approach, sensor streaming data are divided into windows with the equal time slice intervals. This approach is good when dealing with the sensor data that operates continuously in time [6]. In this approach, the challenge is deciding the duration of a time window. If the duration is too small, the sensor sequence in corresponding time window may not be helpful for the machine learning algorithm to take a decision on correct activity. Similarly, if the time window is too long, many activities are placed in one time window and the activity which has the longest time duration will be considered as the one that happened during that interval, while the other activities will be ignored. It may also happen that certain time windows may not have any sensor events. The proposed segmentation method brings an innovation by using the conventional home architecture for segmentation and supports real-time activity recognition with better accuracy.

The remaining part of the paper is structured as follows: Sect. 3 describes the proposed framework of an ambient monitoring system in detail. Section 4 describes experimental analysis, results and performance evaluation. Finally, Sect. 5 concludes the work and gives insight to future work.

3 Proposed Framework of an Ambient Monitoring System

The maturing of sensor technologies and a creation of physical smart environment test beds made possible to create smart environment effectively. Sensors that are deployed in a smart home collect data when resident performs his/her regular activities [8]. When the user performs the activity, motion sensor records the movement in the form of data. Using this data, activity recognition system recognizes the activities that happen in a smart home. Figure 1 shows the proposed framework of an ambient monitoring system. It consists of three phases, namely SH data acquisition, monitoring system and tracking system. In the first phase, data are collected from different sensors like a motion sensor, temperature sensor and door sensor, which are deployed in SH. The second phase is monitoring system. It consists of five subphases.

- Pre-Processing
- Segmentation
- Feature Extraction
- Activity Classification
- Activity Recognition

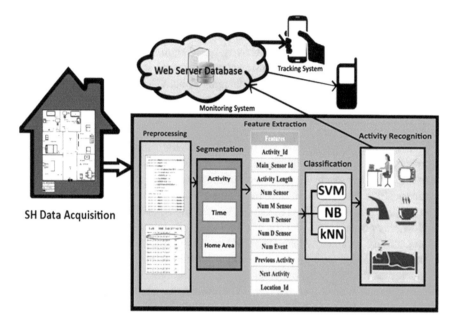

Fig. 1 Ambient monitoring framework

The third phase is the tracking system which consists of a web server database and a mobile application. This phase helps to keep track of the activities happening in a SH using a mobile application. The following subsections discuss the different phases of the ambient monitoring framework.

3.1 SH Data Acquisition

Centre for Advanced Studies in Adaptive Systems (CASAS), a smart home project developed by Washington State University in the United States of America, treats environment as an intelligent agent. The aim of CASAS project is to research on *Helping elders to live in comfort, safety and supportive health using different sensors.* CASAS provides enormous datasets to the researchers that are collected from the real smart home or smart home test bed. In this work, Aruba dataset (House-A) collected from Aruba test bed is used for comparing results. The CASAS project is designed with a smart home kit, viz. Smart Home in a Box (SHiB) [9] which is small in size, easily installable, with understandable infrastructure and ready to use out of the box. This project supports researchers by providing the SHiB kit to the interested researchers to set up a smart home and collects data. In this research, data are collected using CASAS smart home kit $SHiB010$ and collected data are available in a public domain [10].

Date	Time	Sensor_Id	State
04-03-2016	14.17.31	M017	ON
04-03-2016	14.17.31	M019	ON
04-03-2016	14.17.32	M017	OFF
04-03-2016	14.17.33	M019	OFF
04-03-2016	14.17.33	M018	ON
04-03-2016	14.17.33	M019	ON
04-03-2016	14.17.35	M018	OFF
04-03-2016	14.17.36	M019	OFF

Fig. 2 Meal_Preparation activity and corresponding motion sensor events

A real home test bed (House-B) is installed using $SHiB010$ kit. This test bed consists of a kitchen, a bedroom with attached bathroom, an office, a living room and a common bathroom. The test bed smart home is equipped with three temperature sensors, one door sensor and nineteen infrared motion sensors. Each sensor deployed in the test bed has an unique ID assigned to it. The activities happening in the SH is sensed by the sensors as a sensor event and it is acquired wirelessly and stored in the central server. Figure 2 shows an elderly person performing 'Meal_Preparation' activity and the corresponding motion sensor-triggered events.

In order to collect data, 59 healthy participants were identified. Participants were introduced to the layout of the smart home and oriented with the instructions on what they are expected to do. They were instructed with the list of activities, so as to avoid pause or wait before moving to the next activity. There are 11 different activities recorded in the dataset. They are: Bed_to_Toilet, Eating, Enter_Home, Housekeeping, Leave_Home, Meal_Preparation, Relax, Hygiene, Sleeping, Wash_Dishes and Work. Participants were requested to do the activities in their own order of convenience. Executing the activities in the random order was carried out to overcome the possibility of machine learning algorithms from being the sole-deciding factor of activity recognition. The manual annotation of the activities was parallelly done by the experimenter.

Aruba test bed was designed to find the regular activities in a normal life and dataset was collected in the time period 2010–2011. The participant in the test bed was a woman. House-A consists of two BedRooms, a living room, two BathRooms, an office and a Kitchen with Dining area. Totally 40 sensors namely 31 Motion Sensors, 5 Temperature sensors and 4 Door sensors were installed in House-A. There are 11 different activities recorded in the dataset, and this includes Bed_to_Toilet, Eating, Enter_Home, Housekeeping, Leave_Home, Meal_Preparation, Relax, Resperate, Sleeping, Wash_Dishes and Work.

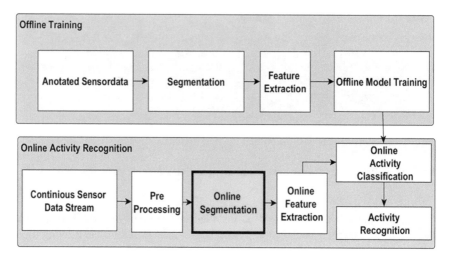

Fig. 3 Proposed monitoring system

3.2 Monitoring System

The monitoring system of the SH is responsible for analysing the data captured by the sensors and identifying the activities. This can be achieved in two ways. First one is offline mode where the predictive model creation which includes pre-processing, feature extraction, activity classification and recognition happens after entire data collection. Second one is online mode where the activity prediction which includes pre-processing, feature extraction, activity classification and recognition happens in parallel during the data collection. In this research, the proposed online segmentation technique recognize the real-time activities. The proposed monitoring system is shown in Fig. 3. Following subsections explain the four phases of monitoring system in detail.

3.2.1 Pre-processing

Data collected from the smart home where the SHiB010 installed is stored as a '.xml' file as shown in Fig. 4. comprising of four tuples $<Date, Time, Target, Value>$. PHP code is written to process this .xml definition to get its related meaning. Pre-processing techniques like data cleaning are applied to make the data error-free.

3.2.2 Segmentation

Continuous flow of sensor data stream is divided into many segments using different approaches is called segmentation. After pre-processing, the sensor sequence is

<publish><channel>rawevents</channel><data><event><by>ZigbeeAgent</by>
<publish><channel>rawevents</channel>
<data><event><by>ZigbeeAgent</by>
        <packagetype>c4:cardaccess_inhome:WMS10-2</packagetype>
        <sensortype>Control4-Motion</sensortype>
        <serial>0006800000021b39</serial>
        <target>LivingRoomAArea</target>
        <epoch>1457082499.461121</epoch>
        <uuid>c5af6f708fbe49ab929864212b5655e3</uuid>
        <message>ON</message>
        <category>entity</category>
</event></data></publish>
<publish><channel>rawevents</channel><data><event><by>ZigbeeAgent</by>
<publish><channel>rawevents</channel><data><event><by>ZigbeeAgent</by>
<publish><channel>rawevents</channel><data><event><by>ZigbeeAgent</by>
<publish><channel>rawevents</channel><data><event><by>ZigbeeAgent</by>
<publish><channel>rawevents</channel><data><event><by>ZigbeeAgent</by>

DATE	TIME	TARGET	VALUE
04-03-2016	14:38:16	M019	09
04-03-2016	14:38:19	M005	ON
04-03-2016	14:38:19	M005	39
04-03-2016	14:38:19	D001	OK
04-03-2016	14:38:21	M005	OFF
04-03-2016	14:38:21	M005	37
04-03-2016	14:38:25	M006	ON
04-03-2016	14:38:25	M006	27
04-03-2016	14:38:26	M006	OFF
04-03-2016	14:38:26	M006	26

Fig. 4 Sensor data stream

segmented equally and classified as an activity by the classifier. This approach is possible with the motion sensor sequence which is collected from the smart home. The motion sensor in the smart home triggers the sequence only when the human activities happen. The triggered sensor sequences are divided into suitable segment or window to recognize the activity. Features are extracted from each segment and used as an instance by various machine learning classifier algorithms. Different segmentation approaches are followed by the researchers to get better activity recognition [5]. In this research, a new online-mode segmentation, *Area-Based Segmentation* is proposed.

In a regular routine, a resident in a smart home transits from one area to another to do regular activities. For example, to cook food the resident has to go to kitchen, for sleeping he has to go to bedroom and so on. This transition helps to do the *Area Based* segmentation. The activity 'Sleeping' happens on the bed in the bedroom. During this activity, only the sensors in the bedroom will trigger the events. When the resident needs to do the activity, Üse_Toilethe should move from bedroom to toilet. During this activity, the sensors in the toilet will trigger the events. When the resident moves from one room (area) to an another, the sensor triggering sequence also will change. This idea is used in the *Area Based* segmentation approach. Sensor streaming data are divided into segments at the point where the sensor in the adjacent area starts triggering. This approach aids in accurately identifying the activities of a single resident.

Activities happening in a smart home are identified with the sensor sequence. Sensor sequence is nothing but a sequence of sensor triggered data. Each sensor in a smart home is assigned with Sensor_ID. Whenever an activity occurs, the related sensors trigger, generate and store related information. For example, during cooking activity the sensors installed in the kitchen M018, M019 are triggered and generate data and store as Date, Time, Senor_ID, State as shown in Fig. 2.

Sensor sequence or sensor-triggered data stream is the series of sensor-triggered events in non-decreasing order of triggered time. $D1 : n = D1, D2, \ldots, Di, \ldots, Dn$ is a sensor-triggered data stream of length n where Di is unique sensor-triggered event. Each individual sensor event consists of Date and Time at which the sensor is

Fig. 5 Association between
home area and activity

triggered, SensorId and State (ON/OFF) of the sensor. The set of activities A happening in a smart home is represented by $A = \{A1, A2, \ldots, An\}$. As discussed earlier, sensor-installed smart home consists of many areas like kitchen, living room and bed room. The smart home H consists a set of areas denoted as $H = \{H1, H2, \ldots, Hm\}$. Each home area $H1, H2, \ldots, Hm$ is installed with a unique set of sensors.

In a sensor-installed smart home, each area H is strongly associated with most of the daily activities. The association between home area and activity is shown in Fig. 5. For example, the area set $H = \{Kitchen, Bathroom, Bedroom, \ldots\}$ is strongly directly associated with the activity set $A = \{Meal\text{-}Preparation, Bed\text{-}to\text{-}Toilet, Sleeping, \ldots\}$ because the activity Meal-Preparation can happen only at kitchen and so on. This strong association helped to propose the concept of area segmentation. Since each home area is installed with unique set of sensors, the sensor-triggered data stream is divided into segments based on the particular home area where the sensor is triggered by an activity. To implement this idea in practical scenario, the concept of change point detection is adopted.

Change point analysis [11] is a method which helps to identify the points where the statistical features change in a sequence of data stream. During activity, the sensors trigger data stream along with corresponding Sensor_ID. In a smart home, sensors are deployed in a specific order. When the activity changes or transition happens from one area to another, the sequence of Sensor_ID also changes. When the activity changes from Sleeping to Use_Toilet or transition happens from bedroom to bathroom, the sensor sequence also changes from M008, M009, M010 to M004, M005. To detect this change point in the data sequence, Pruned Exact Linear Time (PELT) [12] method is used. This algorithm detects optimal change point with a linear computational cost. PELT introduces the pruning which reduces the computational cost with no compromise on perfect change point detection. These change points are the points where the data stream are divided into segments. An algorithm

1. **Input**
2.   Sensor triggered data stream $D_{1:n} = D_1, D_2, D_3, \ldots\ldots, D_n$
3. $D_i=\{Date,\ Time,\ Sensor\_Id,\ State\}$
4.   Home Area set $H=\{H_1,\ H_2,\ldots\ldots,\ H_m\}$
5.   Home Area wise Sensor set
6. $H_1=\{s_{11},\ \ldots\ldots s_{1y}\},\ H_2=\{s_{21},\ \ldots\ldots s_{2y}\},\ldots\ldots\ H_m=\{s_{m1},\ \ldots\ldots s_{my}\}$
7.   Depends on H, y varies
8.   PELT change point detection (equation 1)
9. **Output**
10.   Segmented sensor data stream $S_{1:x} = S_1, S_2,\ldots\ldots\ S_x$
11. **Body**
12. $CP=PELT[D]$
13. $P=1$
14.   **For** $i=1$ to $COUNT(CP)$
15. $S_i=D[P:CP[i]]$
16. $P=P+CP[i]$
17.   **End**

Fig. 6 Segmentation algorithm

Fig. 7 Association between home area and activity

for *Area Based* segmentation using PELT method is represented in Fig. 6. In area segmentation approach, the entire data stream is segmented using PELT change point detection. This segmentation is shown in Fig. 7.

In Fig. 7, the sensor stream data from event 1–7 are segmented as Living Room segment and the event 8–13 are segmented as Office segment and so on. Figure 8

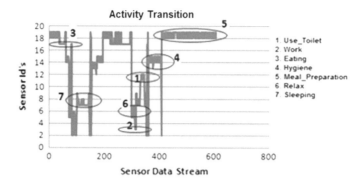

Fig. 8 Relation of sensor data stream and area/activity transition

Table 1 Description of two datasets used for this study

dataset	House-A	House-B
#Sensors	40	23
#Sensor instances	1,048,574	82,285
Activity occurrences	3656	327

shows the graphical representation of changes in sensor data stream with respect to the area/activity transition.

3.2.3 Feature Extraction

The main aim of this work is to analyse sensor data to determine human activity in a smart home environment. Table 1 shows the description of two datasets House-A and House-B.

In the above-mentioned datasets, House-A and House-B have 11 different activities. Among these, only the common and usual activities are considered for this research. Details of the activities and number of occurrences are given in Table 2.

Table 2 Activities and its occurrence in House-A and House-B

Activity	#Occurrence in House-A	#Occurrence in House-B
Bed_to_Toilet	114	11
Eating	185	59
Leave_Home	264	–
Hygiene	–	47
Meal_Preparation	944	59
Relax	1805	52
Sleeping	242	49
Work	112	50
Total activities	3656	327

Table 3 Extracted features used to describe an activity

Feature #	Feature	Description
1	Activity_Id	Id of an Activity
2	Main_Sensor Id	Sensor Id of an Activity
3	Activity Length	Duration of Activity in Seconds
4	Num Sensor	Count of unique sensors that fired during Activity
5	Num M Sensor	Count of Motion sensor reading during Activity
6	Num T Sensor	Count of Temperature sensor reading during Activity
7	Num D Sensor	Count of Door sensor reading during Activity
8	Num Event	Count of sensor trigger in a segment that was generated during Activity
9	Previous Activity	Previously happened Activity in a sequence
10	Next Activity	Next Activity in a sequence
11	Location_Id	House location of an Activity
12	Time_of_Day	24 h divided into four segments. 5 AM–12 PM, 12 PM–17 PM, 17 PM–21 PM, 21 PM–5 PM
13	Day of Week	Day of week SUN-1, MON-2, etc.
14–30	–	Count of each Motion sensor that fired during Activity

In order to find these activities, 30 features are extracted from sensor data and machine learning algorithms are used to recognize the specified activities. Table 3 summarizes the features which are extracted from the raw sensor data.

3.2.4 Classification

Activity recognition models are built using knowledge-driven approach or data-driven approach. The first approach needs prior rich knowledge in respective domain to design activity model using knowledge engineering. Even though this method is semantically clear, it is viewed as static and incomplete, because of the inability to handle uncertainty in temporal data. The second approach has the capability of handling ambiguity and temporal data. In data-driven approach, activity recognition models are built using either supervised, semi-supervised or unsupervised algorithms. Ambient monitoring systems are developed to monitor resident of a smart home. Since activity recognition is purely dependent on resident behaviour, supervised algorithms also known as classifiers are well suitable for ambient activity recognition in smart home. In this study, different classifiers are used to analyse activity accuracy. Out of that, four well-known machine learning classifiers, viz. support vector machine (SVM) kernel techniques which includes SVM polynomial kernel (SVM-P), SVM Gaussian kernel (SVM-R), Naive Bayes classifier (NBC) and k-nearest neighbour (kNN) were responded well. Support vector machine classifier finds the class margin between two classes to classify them linearly. In some real practical cases, linear separable is not useful. So the optimal hyperplane can be constructed and solved using kernel functions. Commonly used kernel functions are polynomial kernel and Gaussian kernel. Though SVMs were originally developed for two-class classification, it can also be applied to multi-class classification by One Vs. All (OVA) and One Vs. one (OVO). OVA forms N binary classifiers for N-class problem by considering ith class with positive labels and others with negative. OVO forms $N(N-1)/2$ binary classifiers and correct class is found by the voting system. OVO is faster and more memory efficient than OVA since each classifier is much smaller [13] and hence in this study, OVO approach is used. The NBC [14] classifier extracts features using relative frequency of features and frequency of activity label in a data stream to understand the association between features F and activity label A using the below mentioned Eq. (1).

$$argmax_{A \in AS} = \frac{(P(F)|P(A))}{(P(F))} \tag{1}$$

kNN [15] is a non-parametric classifier that uses an integer factor K where K is the number of neighbours. This algorithm finds K nearest training data points for a given input a, and predicts the class of a, based on the class of the K data points.

3.3   Tracking System

The smartphone is an ubiquitous device which integrates different technologies and applications. Nowadays, different models of a smartphone with useful features like GPS, the Internet and touch screen are available in the market. These features have made smartphones a necessity than a luxury in human life. This aspect of smartphone has motivated us to develop a mobile app to get an update of every activity that happened in the SH. The intelligent system updates the recognized activity into the webserver database. The smartphone application helps to get the updated activity for the user. Usually, an elder person lives in the SH alone and the caretaker/relatives who lives away from SH, need to monitor the ability of an elder person frequently. In a sensor-deployed SH environment, the collected sensor data are used to recognize the ability with the help of machine learning classifiers. The mobile application is developed for both caretaker and elder. This application updates caretaker with the recognized activities and also provides an option to elders to raise an alarm or send SMS when they need assistance. It also aids in locating the places like ATMs and hospital to get their essential needs done when they are going out alone.

4   Results and Discussion

The proposed framework classifies the daily activity of a human in a smart home environment. Two datasets from CASAS smart home project are used to test the model and verify the classification result. Four well-known classifiers were recognized as

SVM-R	Bed_To_Toilet	Work	Eating	Hygiene	Meal_Preparation	Relax	Sleeping
Bed_To_Toilet	100	0	0	0	0	0	0
Work	0	95	0	0	0	4	0
Eating	0	0	100	0	0	0	0
Hygiene	0	0	0	100	2	0	0
Meal_Preparation	0	0	0	0	98	0	0
Relax	0	5	0	0	0	96	0
Sleeping	0	0	0	0	0	0	100

SVM-P	Bed_To_Toilet	Work	Eating	Hygiene	Meal_Preparation	Relax	Sleeping
Bed_To_Toilet	80	0	0	0	0	0	0
Work	0	80	0	0	0	8	0
Eating	0	0	70	0	2	0	0
Hygiene	0	0	0	100	0	0	0
Meal_Preparation	0	0	20	0	96	0	0
Relax	0	20	10	0	0	92	5
Sleeping	20	0	0	0	0	0	95

NBC	Bed_To_Toilet	Work	Eating	Hygiene	Meal_Preparation	Relax	Sleeping
Bed_To_Toilet	100	0	0	0	0	0	0
Work	0	100	0	2	2	44	2
Eating	0	0	90	0	16	0	0
Hygiene	0	0	0	98	4	0	0
Meal_Preparation	0	0	0	0	78	0	0
Relax	0	0	0	0	0	56	0
Sleeping	0	0	0	0	0	0	98

kNN	Bed_To_Toilet	Work	Eating	Hygiene	Meal_Preparation	Relax	Sleeping
Bed_To_Toilet	60	0	0	2	4	8	2
Work	0	70	0	0	0	4	7
Eating	0	0	60	0	4	0	0
Hygiene	20	0	0	94	0	4	5
Meal_Preparation	0	5	30	2	84	0	0
Relax	20	10	10	0	4	64	5
Sleeping	0	15	0	2	2	20	81

Fig. 9 Confusion matrix of four classifiers using House-B dataset

a benchmark for this study, namely SVM (R-package 'e1071') kernel techniques, namely SVM polynomial kernel (SVM-P), SVM Gaussian kernel (SVM-R), NBC (R-package 'e1071') and kNN (R-package 'class'). These classification algorithms were available in R, an open-source programming language which was developed at Bell Laboratories. The intelligent system stores this recognized activity in the webserver database.

In this research, area dependency segmentation technique is implemented using two datasets. *Activity Based* segmentation and *Time Based* segmentation are also implemented to compare the accuracy of results. In order to see the performance of the segmentation techniques, three classifiers SVM (polynomial, Radial), Naive Bayes and kNN are used and to improve the accuracy of classification, 5-fold cross-validation is used. This validation divides data into 5 sets of size/5, train on 4 datasets and test on 1, and repeat the procedure 5 times and take a mean value. Figure 9 shows the confusion matrix of all classifiers for the test bed (House-B) dataset, using *Area_Based* Segmentation. The rows specify the percentage of predicted activities and columns specify the percentage of actual activities. The value in row I and column J represents the percentage of times an activity J classified as an activity I.

The comparison of the accuracy of all seven activities for four classifiers using House-B dataset by applying *Area Based* segmentation is shown in Fig. 10. It can be observed that the SVM-R and NB classifiers achieved a higher accuracy for Bed_to_Toilet, Work, Eating, Hygiene and Sleeping. The layout of the house, annotation method and classifiers are influencing the performance of the result.

In order to evaluate the proposed segmentation, the performance measures Recall, Precision, F1Score, Accuracy, Kappa value are calculated using False Negatives (FN), False Positives and True positives (TP) [16]. Recall is the measure, that indicates the proportion of positives that are correctly classified and it is specified by Eq. (2).

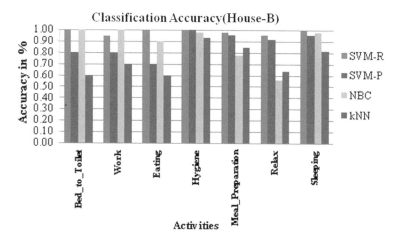

Fig. 10 Comparison of accuracy of all activities for four classifiers using House-B dataset

$$Recall = \frac{TP}{(TP + FN)} \tag{2}$$

The denominator of the above formula is the sum of all positive classifications, apart from a right prediction by the model. Recall measures the ratio of correct prediction of positive class. The Precision measures the ratio of correct positive prediction and it is calculated using Eq. (3).

$$Precision = \frac{TP}{(TP + FP)} \tag{3}$$

The denominator of Eq. (4) is the sum of all positive predictions, so it specifies correctly predicted class. $F1Score$ or $F\text{-}measure$ is a measure of a statistic test's accuracy. $F1score$ is measured as a weighted average of precision and recall and it is calculated using Eq. (4).

$$F1score = \frac{2 * Precision * Recall}{Precision + Recall} \tag{4}$$

F1score measure lies between 0 and 1. Best F1score is closer or equal to 1, and worst F1score is closer or equal to 0. The Accuracy of an activity recognition of all activity instances is calculated using Eq. (5). In this equation, the total number of activity instances are specified by N.

$$Accuracy = \frac{TP}{N} \tag{5}$$

The Kappa statistic or Cohen's kappa coefficient is used to evaluate inter-rate reliability. In Activity recognition, kappa statistics measure the degree of an agreement of the algorithm and actual activity labels. It compares an observed Accuracy A with an expected accuracy E. Equation (6) is used to calculate Kappa statistics.

$$Kappa = \frac{P(A) - P(E)}{P(E)} \qquad (6)$$

Two different approaches are used to recognize an activity in SH environment. First one is offline mode which recognizes the activities after the completion of entire data collection. The second approach, viz. online mode helps to recognize activities when they happen. Segmentation is the earlier step of an activity recognition. *ActivityBased* segmentation is a good example for offline-mode segmentation. In this approach, the starting and end point of the segment is identified using different techniques. During data collection, manual intervention like asking subject to stand still for few seconds, doing activity by following instructions from experts and the smartphone applications are used between transition of two activities to find starting and end point of activities. This manual intervention is used for data stream segmentation and this causes each segment to be directly related to its corresponding activity. In *ActivityBased* segmentation approach, the subjects of the SH environment execute the tasks in a predefined order. Even though, activity recognition using this approach provides better accuracy, the results of this segmentation approach are predictable in nature. In *AreaBased* segmentation, the starting and end point of segmentation is automatically found using the transition of a person from one home area to other. This segmentation is not related to the activity directly. The comparison of the three approaches, viz. *ActivityBased* segmentation, *TimeBased* segmentation and *AreaBased* segmentation with four machine learning algorithms SVM-R, SVM-P, NB and kNN for two data sets, House-A and House-B are shown in Table 4. Accuracy result of *AreaBased* segmentation using SVM-R classifier is comparable with activity segmentation result for both houses. Overall *ActivityBased* segmentation shows better performance for all evaluation measures. Even though, it gives better result, by following activity-defined approach, it is not suitable for real-time activity recognition.

TimeBased segmentation is a good example for online-mode segmentation. The starting and end point of the segmentation is found by dividing data stream into equal time duration. Data segments generated using *TimeBased* segmentation are not directly related to its corresponding activity. So this approach is suitable for online segmentation. Fixing duration for segmentation is a big challenge. In this research, a new online-mode segmentation, viz. *AreaBased* segmentation is proposed. This segmentation is based on the smart home areas like kitchen and bedroom.

Both *TimeBased* and *AreaBased* techniques are suitable for real time activity recognition since the segmentation does not happen based on the happening of the activity. *AreaBased* segmentation gives better result than *TimeBased* segmentation. The high value of F1 score confirms the improved performance in both precision and recall compared to *TimeBased* segmentation approach. In House-B dataset, *AreaBased* segmentation approach, SVM-R classifier achieved F1 score of 0.98 which is 0.06% higher than *TimeBased* segmentation approach. Analysis of these

Table 4 Performance measures of four classifiers

Segmentation approach	Classifier	Precision (%) of house		Recall (%) of house		F1score [0,1] of house		Accuracy (%) of house		Kappa (%) of house	
		B	A	B	A	B	A	B	A	B	A
Activity-based segmentation	SVM-R	99.52	99.69	99.52	98.68	0.99	0.99	99.39	99.11	99.27	98.82
	SVM-P	99.23	99.02	99.27	96.58	0.99	0.98	99.08	98.04	98.91	97.4
	NB	86.83	63.72	85.51	59.37	0.82	0.45	83.79	42.17	80.79	34.81
	kNN	83.31	94.03	83.5	87.95	0.83	0.9	83.79	92.88	80.76	90.44
Time-based segmentation	SVM-R	94.19	98.26	89.71	90.46	0.92	0.94	92.56	94.81	90.63	91.29
	SVM-P	95.49	64.99	94.35	59.37	0.95	0.62	94.02	91.23	92.48	85.14
	NB	76.24	65.15	70.53	83.27	0.73	0.68	82.3	92.96	77.66	88.83
	kNN	83.34	78.47	69.86	62.46	0.71	0.68	82.07	94.87	77.39	91.31
Area-based segmentation	SVM-R	98.42	95.36	98.4	74.09	0.98	0.81	98.46	91.02	98.1	79.33
	SVM-P	92.02	88	87.56	44.28	0.89	0.82	92.82	84.6	91.12	62.09
	NB	87.03	33.97	88.47	36.78	0.85	0.04	87.65	37	84.98	22.44
	kNN	66.48	78.26	66.8	74.66	0.66	0.75	76.92	90.27	71.44	78.38

scores suggests that *AreaBased* segmentation performs better than *TimeBased* segmentation in the accurate identification of activity instance. The overall accuracy of *AreaBased* segmentation using SVM-R of House-B is 98.46% which is 5.9% more than the *TimeBased* segmentation. In summary, the activity recognition accuracy result of *AreaBased* segmentation gives better performance for online activity recognition.

5 Conclusions

Smart home-assisted living facilitates elders to live alone without any assistance. Activity recognition provides the security envisages in a smart home-assisted living. The real smart home test bed is installed using smart home kit *SHiB*010 from CASAS project at Washington State University. A framework for an ambient-assisted living system is proposed in this work. It recognizes the regular activities that happen in a smart home environment. A mobile application is developed to get the update of the activities that happened in the smart home from a remote location. As a future work, this proposed framework will be used to implement real-time ambient monitoring system for elders who need monitoring but lives alone in a SH.

Acknowledgements We would like to thank and acknowledge all the participants who have participated in data collection.

References

1. Mahajan, A., Ray, A.: The Indian elder: factors affecting geriatric care in India. Glob. J. Med. Public Health **2**(4), 1–5 (2013)
2. United Nations, Department of Economic and Social Affairs, Population Division, World Population Prospects The 2015 Revision, July 2015. http://www.un.org/en/development/desa/publications/world-population-prospects-2015-revision.html
3. Kavitha, R., Nasira, G.M., Nachamai, M.: Smart home systems using wireless sensor network. A comparative analysis. Int. J. Comput. Eng. Technol. (IJCET) (2012)
4. Medicinet. http://www.medicinenet.com/script/main/art.asp?articlekey=2152
5. Ni, Q., Belen, A., Pau, I.: The Elderly's independent living in smart homes: a characterization of activities and sensing infrastructure survey to facilitate services development. Sensors (2015)
6. Krishnan, N.C., Cook, D.J.: Activity recognition on streaming sensor data. Pervasive Mob. Comput. **10**, 138–154 (2012)
7. Debes, C., Merentitis, A., Sukhanov, S., Maria, N., Frangiadakis, N., Bauer, A.: Monitoring activities of daily living in smart homes understanding human behavior. IEEE Signal Process. Mag. 33(2) (2016)
8. Cook, D.J., Schmitter-Edgecombe, M., Dawadi, P.: Analysing activity behavior and movement in a naturalistic environment using smart home techniques. IEEE J. Biomed. Health Inform. **19**, 1882–1892 (2015)
9. Cook, D.J., Crandall, A.S., Thomas, B.L., Krishnan, N.C.: CASAS: a smart home in a box. Computer **46**(7), 62–69 (2013)
10. WSU CASAS Dataset. http://casas.wsu.edu/datasets/. Accessed 29 July 2016

11. Aminikhanghahi, S., Cook, D.J.: Using change point detection to automate daily activity seg-mentation. In: 13th Workshop on Context and Activity Modeling and Recognition (2017)
12. Killick, R., Fearnhead, P., Eckley, I.A.: Optimal detection of changepoints with a linear com-putational cost. J. Stat. Softw. 1–19 (2012)
13. Galar, M., Fernndez, A., Barrenechea, E., Bustince, H., Herrera, F.: An overview of ensemble methods for binary classifiers in multi-class problems: experimental study on One-vs-One and One-vs-All schemes. Pattern Recogn. 1761–1776 (2011)
14. Cook, D.J., Krishnan, N.C., Rashidi, P.: Activity discovery and activity recognition: a new partnership. IEEE Trans. Cybern. **43**(3), 820–828 (2013)
15. Li, T., Zhang, C., Ogihara, M.: A comparative study of feature selection and multiclass clas-sification methods for tissue classification based on gene expression. Bioinformatics **20**(15), 2429–2437 (2004)
16. Elhoushi, M., Georgy, J., Noureldin, A., Korenberg, M.J.: A survey on approaches of motion mode recognition using sensors. IEEE Trans. Intell. Transp. Syst. **18**(7), 1662–1686 (2017)

The Conceptual Approach of System for Automatic Vehicle Accident Detection and Searching for Life Signs of Casualties

Anna Lupinska-Dubicka, Marek Tabedzki, Marcin Adamski,
Mariusz Rybnik, Miroslaw Omieljanowicz, Andrzej Omieljanowicz,
Maciej Szymkowski, Marek Gruszewski, Adam Klimowicz, Grzegorz Rubin
and Khalid Saeed

Abstract The European eSafety initiative aims to improve the safety and efficiency of road transport. The main element of eSafety is the pan-European eCall project—an in-vehicle system which idea is to inform about road collisions or serious accidents. As estimated by the European Commission, the implemented system will reduce services' response time by 40%. This will save 2,500 people a year. In 2015, the European Parliament adopted the legislation that from the end of March 2018 all new cars from the EU should be equipped with the eCall system. The limitation of this idea is that only small part of cars driven in UE are brand new (about 3.7% brand new cars were sold in 2015). This paper presents the first concept of an onboard eCall device which can be installed at the owners' request in used vehicles. The proposed system will be able to detect a road accident, indicate the number of vehicle's occupants, report their vital functions, and send that information to dedicated emergency services via duplex communication channel.

1 Introduction

According to the European Commission (EC) estimations, approximately 25,500 people lost their lives on EU roads in 2016 and a further 135,000 people were seriously injured [1]. Studies have shown that thanks to immediate information about the

A. Lupinska-Dubicka (✉) · M. Tabedzki · M. Adamski · M. Omieljanowicz
M. Szymkowski · M. Gruszewski · A. Klimowicz · K. Saeed
Faculty of Computer Science, Bialystok University of Technology, Bialystok, Poland
e-mail: a.lupinska@pb.edu.pl

M. Rybnik
Faculty of Mathematics and Informatics, University of Bialystok, Bialystok, Poland

A. Omieljanowicz
Faculty of Mechanical Engineering, Department of Automatic Control and Robotics, Bialystok University of Technology, Bialystok, Poland

G. Rubin
Faculty of Computer and Food Science, Lomza State University of Applied Sciences, Lomza, Poland

© Springer Nature Singapore Pte Ltd. 2019
R. Chaki et al. (eds.), *Advanced Computing and Systems for Security*,
Advances in Intelligent Systems and Computing 883,
https://doi.org/10.1007/978-981-13-3702-4_5

location of a car accident, the response time of emergency services can be reduced by 50% in rural areas and 60% in urban areas. Within the European Union, this can lead to saving 2,500 people a year [2, 3].

eCall is an initiative with the purpose to bring rapid assistance to people involved in a collision anywhere in the European Union. In case of an accident, an eCall-equipped car will automatically contact the nearest emergency center. The operator will be able to decide which rescue services should immediately intervene at an accident scene. To make such decision, the operator should collect as much information as possible about the causes and effects of the accident and about the number of vehicle occupants and their health condition.

In this paper, the first concept of compact eCall-compliant in-vehicle system is presented, and it can be installed as an additional unit to the car. It detects the human presence, indicates the number of occupants, detects an accident, and reports the state of occupants' vital functions. The proposed device has to be independent of car security systems.

The rest of the paper is organized as follows: In Sect. 2, the authors shortly describe the eCall system with its goals and requirements. In Sect. 3, the authors propose a preliminary approach. To achieve this, the existing sensors and algorithms for accident detection (Sect. 4) and occupants' detection (Sect. 5) and monitoring vital human signs (Sect. 6) are presented. The classification performance of sensors and usefulness is discussed. Finally, the conclusions and future work are given.

2 eCall System

Studies have found that getting immediate information about an accident and pinpointing the exact location of the crash site can cut emergency services' response time by 50% in rural and 60% in urban areas, leading to 2,500 lives saved per year across the European Union [2]. The eCall system, a pan-European emergency notification system, is expected to reduce the number of fatalities in the union as well as the severity of injuries caused by road accidents, thanks to the early alerting of the emergency services. On April 28, 2015, the European Parliament adopted the legislation on eCall type-approval requirements and made it mandatory for all new models of cars to be equipped with eCall technology from March 31, 2018, onward.

eCall is an in-vehicle road safety system which idea is to inform about road collision or serious accident [3]. Since eCall is a standard voice call, it can be triggered manually by vehicle driver or passenger. In case if the vehicle occupants are unconscious, eCall system is able to automatically contact the nearest public-safety answering point (PSAP), operating within the system of the pan-European 112 emergency network. After establishing the connection, eCall system transmits a certain package of basic information about vehicle and accident location (160 bytes). This package, called minimum set of data (MSD) [4], contains among others: latitude and longitude of the vehicle, the triggering mode (automatic or manual), and the vehicle identification number and other information. The purpose is to enable the emergency

Table 1 Content and format of minimal set of data (MSD) (EN 15722)

Block no.	Name	Description
1	Format version	MSD format version set to 1 to discriminate from later MSD formats
2	Message identifier	Incremented with every retransmission
3	Control	Automatic or manual activation, position trust indicator, vehicle class
4	Vehicle ID	VIN according to ISO 3779
5	Fuel type	Gasoline, diesel, etc.
6	Time stamp	Time stamp of incident event
7	Vehicle location	Position latitude/longitude (ISO 6709)
8	Vehicle direction	2° degrees steps
9	Recent vehicle location n-1	Latitude/longitude data
10	Recent vehicle location n-2	Latitude/longitude data
11	No. of occupants	Minimum known number of fastened seat belts omitted if no information is available
12	Optional additional data	For example, passenger data

response teams to quickly locate and provide medical and lifesaving assistance to the accident victims. The MSD format has been specified in detail in the EN 15722 standard (see Table 1).

The eCall is a standard voice call, which enables the vehicle occupants to provide the emergency services with additional details of the accident. However, in some situations, when car's occupants are not able to speak, information about the causes and effects of the accident, especially about the number of the people injured and their state of health, must be deduced by the operator from the additional information received with the MSD package, previously gathered by specialized sensors installed inside the car. Such sensors may include PIR sensors, cameras, seat load sensors, and sensors of vital functions of vehicle occupants.

3 Authors' Approach

The conclusion that can be drawn from the statistics is clear—the earliest notification of the emergency services of the event determines the survivability of the accident. The solution is eCall system. Not every road user, however, wants or affords a new car that would be equipped with it. Hence, the authors of the paper have designed a compact system that could be installed in any vehicle. This would allow every car user to rely on the extra security that it provides, for a relatively small price. This chapter presents a general description of the device's operating concept and design. The scheme of the system is depicted in Fig. 1.

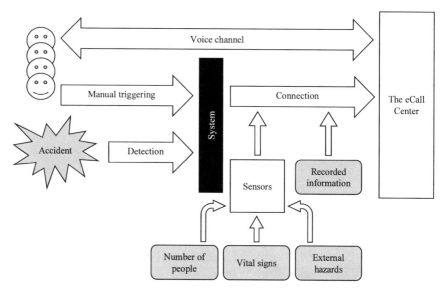

Fig. 1 System block diagram

The first and most important module is the accident detection module. Its task is to detect a dangerous event and launch the entire notification procedure for the relevant service. The key problem here is determining how the accident is defined and how it can be identified (the devices and parameters used). The system, as requested, should also allow manual triggering. This can be used in the following situations: (1) The accident is not serious enough to trigger the sensors, (2) there is no collision, but occupants gravely need help, and (3) the occupants are witnessing an accident and want to inform the eCall services about it.

Another module is the communication module that sends the call to the center. Upon request, additional information recorded in the system (e.g., vehicle data) or read by the sensors (e.g., vehicle location and driving direction read from the GPS receiver) is sent to the center. In addition, the device should be able to establish a voice call with the PSAP operators, allowing them to contact the victims.

The other modules of the system are designed to recognize the situation inside the vehicle to send not only event information to the eCall center, but also assessment of the situation. The first of these modules will detect the presence and the number of people in the vehicle. The task is actually performed before the event occurs—counting when the occupants occupy or leave the seats—to know how many people were in the vehicle at the time of the accident. In addition, post-event monitoring may be provided to inform eCall operators when occupants have left the vehicle (or, e.g., been thrown out if a collision has occurred and they have not fastened their seat belts). Further work will examine the practicality of such a solution.

A separate, but also important, task will be to identify the vital signs of occupants after an accident. If it is possible to assess whether the occupants are alive or unconscious, it will be valuable information for the services preparing to send help. The authors consider a number of methods for evaluating the state of life and a number of sensors used for this purpose, paying attention to the possibility of using them in the vehicle.

In addition to information on the number and severity of injuries, it is possible to collect and share a number of additional information from internal and external sensors—this will be explored and considered in further work. Any details about the condition of the vehicle, the situation inside and also outside (e.g., traffic, external hazards), can be very valuable. All of these additional modules will need to be tested and checked for possibility of use in the intended device—reliability, cost, performance, and ease of installation are the keys. This imposes restrictions on usable sensors. The proposed system assumes that there is no need to use data from the onboard computer of the car, but only sensors are permanently mounted in the device or installed in the vehicle itself.

The project described in this paper assumes yet another kind of information sent to the eCall center—these are all kinds of data input by vehicle users that aim to better and more accurately inform eCall and rescue services. This may include information about illnesses, pregnancy, children's presence, and blood type. Providing this information would not be mandatory, but system users would be given the opportunity if they think this could improve their safety.

The proposed system will therefore include: a GPS vehicle positioning system, set of sensors for accident detection, a digital infrared camera (or camera system) for detecting vehicle occupants' presence, and a set of sensors for analyzing passengers' vital functions. In the opinion of the authors, in order to increase the reliability of the information provided by the system, the individual sensors should be multiplied. The activity diagram of the system is presented in Fig. 2. The main function of the system will be continuous reading of sensor data and analysis of whether the current values of the measured signals do not exceed the acceptable signal levels. In this case, the system will trigger the automatic accident notification. Simultaneously, the system continually processes the data that will be used to generate the MSD package.

Recently, a number of systems with purposes to detect vehicle accidents are described: [5–9]. These approaches generally tend to use accelerometers and GPS tracking devices, especially embedded in modern smartphones.

4 Accident Detection

The key element of the described system is a mechanism which determines whether an accident has occurred or not. The mechanism turns on other components of eCall system and also initializes the rescue operation. It should conclude to high reliability of the whole structure, i.e., no false notifications sent.

However, the most crucial task of the whole system is to protect human health and life; therefore, system is allowed to notify public-safety answering point (PSAP) in case when accident was minor. By accident classified as minor, eCall establishes

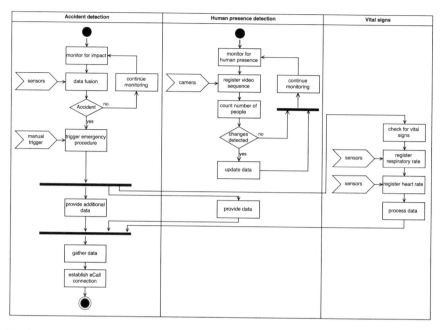

Fig. 2 Activity diagram of the system

voice connection [2] between PSAP staff and accident participant. Based on the voice conversation, PSAP operator determines if rescue operation must be conducted or not. However, fundamentally, data transfer in eCall system must be automatic. Therefore, accident detection mechanism is the obligatory part of the system. It consists of two parts: hardware which is broadly understood motion sensor and software which is data analysis algorithm and also decision algorithm.

Authors of this article decided that their mechanism can work independently from onboard systems of the car. That particular approach creates possibility to equip with that mechanism cars of a different age (even very old), not only those newest ones which are fitted up with advanced sensors. Such versatility complicates selection of a hardware platform and complexity of algorithms because it is crucial to note that accident detection decision must be made without knowledge from onboard systems of the car such as belt fastening or air bag release status.

According to the above, the authors decided to consider to use a vast range of sensors and data fusion algorithms. As a source of the information, the following types of devices and sensors are selected: accelerometers, gyroscopes, pressure sensors, temperature sensors, and sound detectors. Accelerometers are the most common components of collision detection systems [10]. They respond to sudden/rapid changes in the rate of movement (acceleration and/or delay) by measuring them in units g's—the magnitude of the acceleration. It was assumed that they would be placed in several places of the vehicle. The use of a single element can generate unwanted activation of an alarm signal caused by, for example, a point strike near a sensor or a sudden strong shake of a vehicle when it got on an obstacle on the road (hole or bump). Accelerometers are useful to determine if the vehicle is struck by a permanent obstacle or another vehicle. But there are many possible accident scenarios [11]. These include the situation of overturning and/or roofing (rollover) of the vehicle. Delays occurring in such cases are considerably less than on impact, and the decision based on the data provided by accelerometers can be false (the accident would be not noticed). The proper sensor in such situation is the gyroscope [12], which identifies the phenomenon of rotation of the vehicle in each of the axes of the classical reference system x, y, z.

In most publications [13, 14], accident detection systems are based on the use of the accelerometers and gyroscopes. Due to the diversity of accident situations, in practice unpredictable, the authors have decided to use additional pressure sensors and sound detectors. Pressure sensors are useful during rapid increase of pressure in the cabin. This phenomenon is accompanied by explosions. Such situation may occur when accident was not serious (neither high values of delays nor the rotation of the vehicle appears); however, fuel tank might be damaged and fuel may have been ignited. That might lead to an explosion that pressure sensors and temperature sensors easily detect. Car structure deformation is the result of a vast number of car accidents. Detection of this deformation can be performed by a specialized sound detector aimed at detecting the specific sound of crashed metal.

All above-mentioned sensors only gather information which later is used to determine if accident happened or not. Sensors are not programmed to invoke alarm notification to the PSAP. It is crucial to develop decision mechanism, which can analyze information gathered with the use of sensors and generate information (MSD) about accident which is sent to PSAP. Such algorithms are used in newer cars for air bags and air curtain systems; however, they cannot be directly used for eCall system. It is because they are programmed to be used before accident [15] (to prevent humans from being hurt), but eCall demands information after. The objective of the proposed compact eCall-compliant device is to be used after accident. The simplest way to use sensor data is to set threshold values above which it is possible to determine if an accident has occurred. However, vast number of articles points out that such method is inaccurate. The authors undertook the task of creating accurate data analysis and decision algorithms which will use data from different types of sensors with a use of data fusion techniques [16]. Expected outcome are algorithms which can be used to determine with a high accuracy if accident occurred in immense number of circumstances.

5 Human Presence Detection

As a part of the proposed in-vehicle system, it is required to detect the number of occupants and their vital condition. In this subsection, the literature review of the solutions connected with the human presence detection and our idea how to solve this problem are presented.

There are many different approaches to human detection problem, but small number of them is associated with the vehicles. An interesting idea was presented in [17]. The authors of this solution proposed the system that detects moving objects in a complex background. In this paper, Gaussian mixture model (GMM) was claimed as an effective way to extract moving objects from a video sequence. The authors also claimed that conventional usage of GMM suffers from false motion detection in complex background and slow convergence. To achieve robust and accurate extraction of moving objects, a hybrid method by which noise is removed from images was used. At the first stage, the proposed model consists of fourth-order partial differential equation (PDE). At the second stage, usage of relaxed median filter on previously obtained output is claimed. The authors provided experimental results in which it is clearly visible that their approach has detection rate equal to 98.61% what compared to conventional usage of GMM is more convenient, forasmuch in this paper GMM was claimed to have 73.5% accuracy.

The authors of [18] claimed that technology development enabled usage of human detection methods in the intelligent transportation system in smart environment. In this paper, it was pointed out that it is a huge challenge to implement an algorithm that will be robust and fast and could be used in Internet of things (IoT) environment. In the case of these systems, low computational complexity is crucial due to available resources. The authors proposed to use histogram of oriented gradient (HOG) method. Conventional usage of this algorithm gives accuracy on satisfactory level, although it is really expensive to compute. The described solution aims to reduce the computations using approximation methods and adapts for varying scale. In this work, it was pointed out that experiments were done on online available datasets like the one that is shared by Caltech. This approach uses also specific hardware solutions, and it is based on FPGA structures. In the paper, experimental results were presented. In this case, when the database has grown to 1000 samples, accuracy level is equal to 95.57% and when the database has grown to 2000 samples, accuracy level has decreased to 86.43%.

Both the recently proposed solutions are not connected with the vehicles, but it is considered that it is possible to adjust them to use in the car environment. Another concept that the authors would like to present is the one described in [19]. This article provided an idea to create the system for people counting in public transportation vehicles. This concept was created due to the problem of monitoring the number of occupants getting in or out of public transportations in order to improve vehicle's door control. In this work, both hardware and software implementations are combined. This approach uses stereo video system that acquires a pair of images. In the next steps, disparity image computation and 3D filtering are claimed to be used. In this article, no experimental results were provided; however, it is claimed that the system was installed and tested in natural environment in the bus.

An interesting approach was also described in [20]. In this case, people were not counted but the authors take into consideration automatic traffic counting. Aim of the experiments was to define the requirements to be applied to traffic technologies to match specific applications. The authors tested seven different traffic monitoring systems. In the paper, it is claimed that the study lasted one year. As a result, the

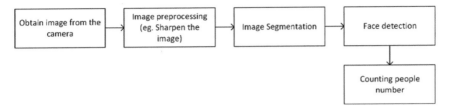

Fig. 3 Algorithm block diagram of people counting

specific test procedures were developed and the most reliable vehicle counting system was chosen. The authors of [20] inform that vehicles were counted manually and then the result was compared with automatic analysis of the recorded video.

To recapitulate, the authors would like to note the possibility of using various image sources, among other standards, infrared and thermal vision. It is important to stress various kinds of nature of visual data for different cameras; for example, watershed algorithm [20] could be used for thermal image segmentation. Individual path of image seems to be, however, rather standard in computer vision application, acquisition, preprocessing, segmentation, and selected algorithms for face detection. Additional possibility with many types of cameras is using different placements in car, in order to increase feasibility. As a final note: face detection is only a subtask of the proposed solution, ready-made implementations, like OpenCV [21] might be used as well.

On the basis of performed experiments and the literature review, authors would like to present short description of their approach to counting people in a vehicle. Due to the fact that authors' study is in the early phase, only block diagram and short information about each step are presented. No experimental results are provided.

At the beginning, it should be claimed that authors would not like to use sensors that are built in the car. Usage of them is connected with car system breach of integrity. What is more the authors would like to check whether usage of different camera types like webcam, infrared camera, or thermal camera could provide satisfactory accuracy level. Processing algorithms will be connected with the images obtained from used camera. The general concept is presented in Fig. 3.

It should be pointed out that people counting has to be done when the vehicle engine has just been started. This can provide information about number of people that are in the car when the journey has started. Counting them when an accident occurs sometimes could be hard to complete. The first step of the authors' approach is to capture image from the camera installed inside the car. Authors' experiments take into consideration also multiple installed cameras. Usage of them will decrease number of people false non-recognition rate. When the image is available, it should be preprocessed. In this step, different processing algorithms could be used, for instance sharpen or median filter. All operations used in this phase that will provide better image quality are included.

In the third step, authors' idea image segmentation is proposed. In this case, type of segmentation is connected with the selected camera type. For example, when thermal camera is analyzed, watershed segmentation [20] could be used. In the fourth step, presented idea is connected with the face detection. A library providing a multitude of algorithms called OpenCV [21] might be implemented in authors' solution. This framework has ready-made implementations that detects faces in the images. In the final step, authors' algorithm people counting is done. This step is based on the number of detected faces.

6 Vital Signs

This section describes vital signs of occupants' aliveness and health that could be gathered by the proposed compact eCall-compliant device using specialized sensors and reported to PSAP, as well as the original number of occupants detected in the car, as described in Sect. 5. This information is vital to the occupants' well-being, as sometimes serious crashes do negligible damage to human beings, and in reverse.

6.1 Respiratory Rate

Respiratory rate is one of the basic vital signs that can be used to asses individual state. There are several methods to measure respiratory rate, and various types of devices are used in clinical practice and research [22]. Registration devices may be categorized by a physical phenomenon resulting from breathing action that they measure such as: respiratory sounds, respiratory airflow, chest and abdominal movements, CO_2 emission, and heart activity. Devices can be also grouped into two classes: requiring contact and contactless. Devices in the first class require direct contact with subjects' body and are usually attached to his/her skin. Instruments in the second category are capable of measuring respiratory from distance and do not make any contact with the subject. Table 2 presents different types of devices and their categories.

As shown in Table 2, respiration sound can be recorded using microphone located in proximity of subjects' nose or mouth. This usually requires attaching the microphone to subjects' head or neck to obtain stronger audio signal.

Respiratory airflow may be measured as a pressure exerted by exhaled air and also by detecting changes in a temperature of a region located near nose that is alternately heated and cooled during exhale and inhale phrases. Measuring airflow pressure can be conducted using devices attached to head such as nasal cannulae or mask. The temperature variability due to respiratory actions can be registered by both contact (oronasal thermistor) and contactless (thermographic camera) devices.

Chest and abdominal movements allow for most precise respiratory measurements. Typically, this is conducted using belts placed around rib cage or abdominal for chest and abdominal respiratory rate registration. The belt can measure tension or changes in electrical impedance/inductance levels as indicators of respiratory actions. This technique requires direct contact with subjects' body. However, several stud-

Table 2 Types of devices for registration of respiratory rate

Measured phenomena	Type of device	Requires contact
Respiratory sounds	Microphone mounted in proximity of nasal/mouth	Yes
Respiratory airflow	Nasal or oronasal thermistor that registers temperature difference between inhale and exhale air	Yes
	Thermographic camera measuring changes in skin temperature near nasal due to inhale and exhale airflows	No
	Nasal pressure transducer that indirectly measures the volume of exhaled air	Yes
Respiratory chest and abdominal movements	Tension/electrical impedance/electrical inductance sensors that measure the extent of chest/abdominal perimeter changes	Yes
	Visible light or infrared camera that measures chest/abdominal perimeter changes	No
	Impulse radio ultra-wideband and Doppler radars	No
Respiratory CO_2 emission	Transcutaneous CO_2 monitoring through heated electrode applied to skin (e.g., arm)	Yes
Heart activity	Detection of modulation caused by respiration in signal from electrocardiogram (ECG) electrodes placed on chest skin	Yes

ies show that chest and abdominal moments can be also registered remotely using cameras and radar sensors with high accuracy.

Respiratory detection may be also based on diffusion of CO_2 gas to the skin. This technique requires placing heated (42°C) electrode on subjects' skin and allows to estimate breathing activity through monitoring changes in CO_2. Breathing action has also influence on heart activity that can be detected in ECG signal.

Selected Techniques for Respiratory Rate Detection in Car Environment

Most of the techniques presented in Table 2 find its use in clinical practice. However, the assessment of their potential application to detection of car drivers and occupants' state must be conducted separately. Techniques that are preferable should be easy to use and should not require additional actions from driver or occupants. Therefore, special sensors that need to be attached to body, such as ECG and CO_2 electrodes, oronasal thermistor, or microphones, are not best suited to this usage. Another requirement is the ability to work inside vehicle environment. This includes robustness to different sources of noise and lighting conditions, and accurate signal registration given different positions and movements of subjects. Selected devices should also continue to function in case of an accident and be able to register data if the subject is unconscious. The following sections preview selected techniques that may be applied given those requirements.

Radar-based   Radar-based registration allows for precise breathing rate measurement and can be used even in case of occlusion by other objects. In the study [23], the authors used impulse radio ultra-wideband (IR-UWB) radar to monitor respiration and heart rate of a driver. Data from radar was processed in the following stages: preprocessing with loop-back filter to remove clutter, extraction of received signals with vital sign information using maximum likelihood estimation and thresholding based on R^2 statistics, vital signal reconstruction, and final analysis using FFT. The proposed solution allowed to reliably register breathing and heart rate in driving conditions.

Pressure sensor-based   In project HARKEN [24], drivers' heart and respiratory rates were detected using special textile sensors that are sensitive to pressure exerted by subjects' body. The material was embedded in drivers' safety belt and seat and could detect changes of pressure caused by cardio and respiratory movements. The obtained signals had to be filtered due to noise from vibration of moving vehicle and body movement. The authors used adaptive filter and reference noise measured using accelerometer to clear the signal before extraction of respiration and heart rate signals.

Camera-based   Camera-based techniques can be fully automatic and do not require any user iterations. In the literature, many methods have been proposed that can be used to extract breathing signal from video sequence.

One of the approaches to measuring movements resulted from breathing is to apply the optical flow method to selected region of interest (ROI) in a video sequence. In [25], the ROI was rectangular area located on subjects' chest. Its location was determined relative to face position that in turn was detected by means of Viola and Jones algorithm. After localization, vertical motions caused by breathing signal were extracted using median values from optical flow method followed by smoothing filter. The presented approach was tested in laboratory environment with different poses such as upright, tilted, and lying.

Breathing signal can be also detected by tracing movements of shoulder or chest contours that can be extracted from camera images. In the work [26], the vertical movements of shoulders' edge inside selected rectangular region were analyzed. The edge was extracted from differential image computed along vertical direction. The ROI was then divided into two subregions: upper and bottom. The breathing signal was computed as the normalized difference between those two subregions. In order to account for body movements not related to breathing, authors used phase correlation method to compute a shift that was applied to ROI position before estimating breathing rate.

Another approach to breathing rate detection is based on Eulerian video magnification framework [27]. The Eulerian video magnification (EVM) method allows to amplify pixel intensity changes at selected frequencies in video sequence and can be used to magnify small movements related to breathing action. This magnification allows for easier and more precise detection. One of the applications presented in [27] was the magnification of respiratory movements in video sequence of infant child. Other examples of EVM application to breathing rate estimation can be found in the works [28, 29]

Most of the presented methods utilized cameras in visible light spectrum. This may be a major drawback in low-light conditions, for example when driving at night. One solution to this problem is to use infrared cameras that illuminate recorded scene with infrared light (700–1000 nm spectral range). Infrared light is not visible to human eye and will not disturb car drivers and other road users. Techniques for image processing and breathing rate detection that may be used with grayscale images can be also applied in this case. For example, in [28] the authors use EVM algorithm with video sequence from IR camera.

Lower frequencies of infrared light captured by thermal cameras allow to record changes of the skin temperature resulting from breathing action. Example of a solution that uses thermal images can be found in the work [30]. The authors use ROI located near nostrils to compute average temperature in this region and based on its changes estimate breathing rate. They also propose ROI tracing algorithm based on particle filter to allow for moving subjects.

The solutions for breathing rate estimation using cameras are usually assessed in laboratory environment, where the subject is sitting in front of a camera. This setting is different from application in the proposed system where the observed subject will be inside vehicle. An example of a study, in which the camera-based solution was evaluated inside a car, can be found in [31]. During experiments, the authors evaluated different camera positions and lighting conditions. The proposed method, after initial preprocessing, performed ROIs localization based on frame differentiation technique. In the next step, after extracting motion signals, non-period signals were discarded and highly correlated semi-periodic signals were averaged to produce unique motion. The final respiration rate was calculated using short-time Fourier transform.

6.2 Heart Rate

Heart rate measurement can be performed using various methods that can be also categorized based on measured phenomena and whether they require direct contact or not [32]. Table 3 summarize techniques and their categorization.

The registration of changes in electrical potential due to heart activity can be carried out using electrical electrodes placed on the subjects' skin in the vicinity of heart. This technique requires direct contact and is relatively cumbersome; however, it provides rich and precise information about heart action.

The change in light absorption cased by volume change due to pulse pressure can be measured using photodiode or phototransistor. The skin is usually illuminated

Table 3 Types of devices for registration of heart rate

Measured phenomena	Type of device	Requires contact
Changes in electric potential	ECG electrodes placed on chest skin (Holter [33])	Yes
Changes in light absorption cased by fluctuations in blood volume	Illumination LED paired with photodiode or phototransistor	No
	Visible light or infrared camera	No
Body movement cased by heartbeat	Impulse radio ultra-wideband and Doppler radars	No
	Tension/electrical, impedance/electrical, inductance/pressure sensors that measure body movement	Yes
Changes in temperature cased by blood flow	Thermographic camera	No
Human speech sound	Microphone	No

using LED. Such technique is used by smartwatches and smartphones that can measure heart rate.

Changes in light absorption can be also captured using camera. Flow of blood causes temperature changes in body, especially in the vicinity of blood arteries. Such changes can be registered using thermographic camera and used to estimate heart rate [28].

Displacement of body due to heartbeat action can be measured with radars. According to many studies, such measurement can provide accurate heart rate values [23].

It has been observed that heart rate can be detected in human speech due to its relation to variations of vocal cord parameters via the larynx, which is indirectly connected to the human circulatory system [34].

Selected Techniques for Heart Rate Detection in Car Environment As it was mentioned earlier, the techniques that may be applicable in car environment should be easy to use, robust to noise, present driving vehicle, and function in case of an accident. In the following sections, we preview selected methods.

Camera-based Changes of light absorption due to blood flow can be registered from video camera in visible light or IR. In the work [28], the authors used EVM framework to magnify changes due to blood flow in IR camera images in selected ROI to measure heart rate. The experiments conducted in laboratory environment showed that obtained measurement was accurate when the distance to the subject did not exceed one meter due to decreasing strength of reflected illumination versus thermal noise emitted by background objects.

Radar-based Heart rate can be accurately estimated by measuring body displacement using UWB radar. The techniques that are used are similar to respiratory rate detection using radar, as it was described earlier, the main difference being the frequency range of registered signal. Description of such approach to estimate heart rate of car drivers can be found in [23].

Photodiode-based Photodiodes installed in devices such as smartwatches or special wristbands measure heart rate by detection changes in light absorption in illuminated skin. If present, such signal can be communicated by wireless technology to processing unit of the eCall system.

Pressure sensor-based Example of heart rate monitoring by means of pressure sensors installed in car safety belt and driver's seat can be found in project HARKEN [24]. The techniques that are used are the same as in respiratory detection and were described in the previous section.

7 Conclusions and Future Works

eCall seems to be an excellent initiative to save lives; however, one must note the limitation of the directive, as new cars sold in EU in 2015 constitute only about 13.7 million versus total of over 270 million cars driven in EU [35, 36] (partially estimated). Thus, the idea of compact and cheap device adjoined to existing cars is surely a very interesting proposal. In this article, the authors have described the eCall solution, mentioned and evaluated a variety of possible algorithms used for such compact eCall-compliant device, and presented the idea of their own approach. The first steps of the suggested methodology have been given with the flowchart of their method. The related subtasks are detection of car accidents, detection of the number of occupants in the vehicle, and estimation of their condition based on vital signs.

Future works in this area are: gathering data from crash simulations, selection of sensors, experimental sensor placement, and development of algorithms for (1) accident detection, (2) human presence detection, and (3) car occupants' vital sign reading. At the moment, the work on the described system is carried out in parallel on two levels: accident detection and analysis of information about vehicle occupants. The accomplished stages of accident detection include obtaining data on acceleration/delays from the crash test, selection of sensors (accelerometer and gyroscope to assist in determining the scale of an accident), and establishing initial thresholds which trigger emergency procedure. The laboratory station design to carry out collision tests has also been developed and is under construction. Works on vehicle occupants' presence detection are in progress. It required collecting data and creating a real photograph data of people in the vehicle. The algorithms for determining the number of passengers are to be selected and configured with optional implementation of mixture of experts' approach.

Acknowledgements This work was supported by grant S/WI/1/2018 and S/WI/2/2018 from Bialystok University of Technology and funded with resources for research by the Ministry of Science and Higher Education in Poland.

References

1. Road Safety: Encouraging results in 2016 call for continued efforts to save lives on EU roads. http://europa.eu/rapid/press-release_IP-17-674_en.htm. Accessed 24 Dec 2017
2. eCall: Time saved = lives saved. https://ec.europa.eu/digital-single-market/en/eCall-time-saved-lives-saved. Accessed 24 Dec 2017
3. European Parliament makes eCall mandatory from 2018. http://www.etsi.org/news-events/news/960-2015-05-european-parliament-makes-ecall-mandatory-from-2018. Accessed 24 Dec 2017
4. ETSI eCall Test Descriptions—ETSI Portal. https://portal.etsi.org/cti/downloads/TestSpecifications/eCall_TestDescriptionsv1_0.pdf. Accessed 24 Dec 2017
5. http://iheero.eu/wp-content/uploads/sites/3/2017/12/I_HeERO-Act2-Webinar-eCall-for-HGV-buses-and-coaches-2017-12-14.pdf. Accessed 24 Dec 2017
6. White, J., Thompson, C., Turner, H., Dougherty, B., Schmidt, D.C.: Wreckwatch: automatic traffic accident detection and notification with smartphones. Mob. Netw. Appl. **16**(3), 285–303 (2011)
7. Zaldivar, J., Calafate, C.T., Cano, J.-C., Manzoni, P.: Providing accident detection in vehicular networks through OBD-II devices and Android-based smartphones. In: Proceedings of the IEEE Conference on Local Computer Networks, pp. 813–819 (2011)
8. Watthanawisuth, N., Lomas, T., Tuantranont, A.: Wireless black box using mems accelerometer and GPS tracking for accidental monitoring of vehicles. In: Proceedings of the IEEE International Conference on Biomedical and Health Informatics, pp. 847–850 (2012)
9. Ahmed, V., Jawarkar, N.P.: Design of low cost versatile microcontroller based system using cell phone for accident detection and prevention. In: 2013 6th International Conference on Emerging Trends in Engineering and Technology (ICETET), pp. 73–77 (2013)
10. Amin, S., et al.: Kalman filtered GPS accelerometer based accident detection and location system: a low-cost approach. Curr. Sci. **106**(11) (2014)
11. Amin, M.S., Sheikh Nasir, S., Reaz, M.B.I., Ali, M.A.M., Chang, T.-G.: Preference and placement of vehicle crash sensors. Tech. Gaz. **21**(4), 889–896 (2014)
12. Classen, J., Frey, J., Kuhlmann, B., Ernst, P.: MEMS gyroscopes for automotive applications. In: Components and Generic Sensor Technologies. Robert Bosch GmbH
13. Islam, M., et al.: Internet of car: accident sensing, indication and safety with alert system. Am. J. Eng. Res. (AJER) **02**(10), 92–99. e-ISSN 2320-0847, p-ISSN 2320-0936
14. Saiprasert, C., et al.: Detection of driving events using sensory data on smartphone. Int. J. ITS Res. **15**, 17–28 (2017). https://doi.org/10.1007/s13177-015-0116-5
15. Kaminski, T., Niezgoda, M., Kruszewski, M.: Collision detection algorithms in the eCall system. J. KONES Powertrain Transport **19**(4) (2012)
16. Castenedo, F.: A review of data fusion techniques. Sci. World J. **2013**, Article ID 704504, 19 (2013)
17. Fazli, S., Pour, H.M., Bouzari, H.: A robust hybrid movement detection method in dynamic background. In: Telecommunications and Information Technology 2009, ECTI-CON 2009, 6th Conference, Pattaya, Chonburi, Thailand Proceedings, (2009)
18. Sageetha, D., Deepa, P.: Efficient scale invariant human detection using histogram of oriented gradients for IoT services. In: 2017 30th International Conference on VLSI Design and 2017 16th International Conference on Embedded Systems Proceedings (2017)
19. Bernini, N., Bombini, L., Buzzoni, M., Cerri, P., Grisleri, P.: An embedded system for counting passengers in public transportation vehicles. In: 2014 IEEE/ASME 10th International Conference on Mechatronic and Embedded Systems and Applications Proceedings (2014)
20. Bellucci, P., Cipriani, E.: Data accuracy on automatic traffic counting: the SMART project results. Eur. Transport Res. Rev. **2**(4), 175–187 (2010)
21. Vanhamel, I., Sahli, H., Pratikakis, I.: Automatic watershed segmentation of color images. Comput. Imaging Vis. **18**, 207–214 (2000)
22. Al-Khalidi, F.Q., Saatchi, R., Burke, D., Elphick, H., Tan, S.: Respiration rate monitoring methods: a review. Pediatr. Pulmonol. **46** (2011)

23. Leem, K.S., Khan, F., Cho, H.S.: Vital sign monitoring and mobile phone usage detection using IR-UWB radar for intended use in car crash prevention. Sensors **17** (2017)
24. HARKEN. http://harken.ibv.org/. Last accesed 1 July 2017
25. Lin, K.Y., Chen, D.Y., Tsai, W.J.: Image-based motion-tolerant remote respiratory rate evaluation. IEEE Sens. J. **16**, 3263–3271 (2016)
26. Shao, D., Yang, Y., Liu, C., Tsow, F., Yu, H., Tao, N.: Noncontact monitoring breathing pattern, exhalation flow rate and pulse transit time. IEEE Trans. Biomed. Eng. **61**, 2760–2767 (2014)
27. Wu, H.-Y., Rubinstein, M., Shih, E., Guttag, J., et al.: Eulerian video magnification for revealing subtle changes in the world. ACM Trans. Graph. **31**, 1–8 (2012)
28. He, X., Goubran, R., Knoefel, F.: IR night vision video-based estimation of heart and respiration rates. In: 2017 IEEE Sensors Applications Symposium (SAS), pp. 1–5 (2017)
29. Al-Naji, A., Chahl, J.: Remote respiratory monitoring system based on developing motion magnification technique. Biomed. Signal Process. Control **29**, 1–10 (2016)
30. Zhen, Z., Jin, F., Pavlidis, I.: Tracking human breath in infrared imaging. In: Fifth IEEE Symposium on Bioinformatics and Bioengineering (BIBE'05), pp. 227–231 (2005)
31. Solaz, J., Laparra-Hernndez, J., Bande, D., Rodrguez, N., Veleff, S., Gerpe, J., et al.: Drowsiness detection based on the analysis of breathing rate obtained from real-time image recognition. Trans. Res. Procedia, **14**, 3867–3876 (2016)
32. Kranjec, J., Begu, S., Gerak, G., Drnovek, J.: Non-contact heart rate and heart rate variability measurements: a review. Biomed. Signal Process. Control **13**, 102–112 (2014)
33. Szczepanski, A., Saeed, K.: A mobile device system for early warning of ECG anomalies. Sensors **14**(6), 11031–11044 (2014)
34. Mesleh, A., Skopin, D., Baglikov, S., Quteishat, A.: Heart rate extraction from vowel speech signals. J. Comput. Sci. Technol. **27**, 1243–1251 (2012)
35. https://www.best-selling-cars.com/europe/2016-full-year-europe-best-selling-car-manufacturers-brands/. Accessed 18 Nov 2017
36. Eurostat—Passenger cars in the EU. http://ec.europa.eu/eurostat/statistics-explained/index.php/Passenger_cars_in_the_EU. Accessed 18 Oct 2017

Disaster Management System Using Vehicular Ad Hoc Networks

Suparna Das Gupta, Sankhayan Choudhury and Rituparna Chaki

Abstract The progress and improvements of distributed networks have played a decisive role in enabling researchers to consider new solutions for various VANET applications such as transportation, road safety, driving assistance, disaster management system, and lots more. Disaster response and management has enabled transport and communications to play an imperative role in dipping loss of life, economic cost, and disruption. VANETs can be measured as a rift of distributed networks. Vehicular ad hoc networks have been implemented as an automated system, and communication has been addressed through vehicular nodes. Since the last two decades, the number of disasters on the side of the road has evidently augmented due to the hasty augment in the number of road vehicles. The damage caused to the human race by disasters has been a severe crisis both national and international arena. In this proposal, we have proposed a scheme disaster management as an application of vehicular ad hoc networks. This proposal would help those affected with an automated system using the functionality of smart vehicles, which would correspond to each other through communication from vehicle to vehicle and could communicate with the drive side of the road through the vehicle–infrastructure communications. Our proposed disaster management system is developed as an application of existing communication system.

Keywords Disaster management · Smart vehicles · Vehicular ad hoc networks
Incentive strategy · Distributed networks

S. Das Gupta (✉)
JIS College of Engineering, Kalyani, West Bengal, India
e-mail: suparnadasguptait@gmail.com

S. Choudhury · R. Chaki
University of Calcutta, Kolkata, West Bengal, India
e-mail: sankhayan@gmail.com

R. Chaki
e-mail: rituchaki@gmail.com

© Springer Nature Singapore Pte Ltd. 2019
R. Chaki et al. (eds.), *Advanced Computing and Systems for Security*,
Advances in Intelligent Systems and Computing 883,
https://doi.org/10.1007/978-981-13-3702-4_6

1 Introduction

Vehicular ad hoc networks have become a popular for both academic application and industry. In [1], authors introduced the notion of vehicle as a resource (VaaR), which has paid throughout in vehicular potential. In order to lessen vehicular mortality and increase the application of ITS and services, vehicles are outfitted with components such as sensors, actuators, electronic control units (ECU) for intra-vehicle communication along with processing and operational control and classified as intelligent vehicles. The included on-board unit (OBU) for interacting with drivers, displaying alerts, issuing alerts that provides automotive or information services can also perform the task of computation supported by profuse storage capacities.

The VANETs consist of vehicles and roadside units (RSUs) as network nodes and inter-vehicle communications (IVCs) together with the roadside-to-vehicle communication (RVC). Vehicles can correspond with nearby vehicles known as communication from vehicle to vehicle (V2V) and also with the side of the road infrastructure also known as vehicle–infrastructure (V2I). IVC and RVC can be divided into two categories: application security-related and application-related information. VANET using vehicle communication between neighboring objects (V2X) would also be taken care of. Benefiting from the large capacities of vehicles, vehicle nodes have long diffusion ranges and virtually unlimited lifetimes.

In this proposal, we have tried to explore different domain VANET applications. As India is a developing country, the number of vehicles on the road increases day by day, causing complications in traffic management. This drastically reduces the life of a road and increases the accident rate. To handle this situation, we have proposed a solution using the communication systems V2V, V2I, and V2X.

The rest of the article is arranged as follows. Related research and comprehensive surveys are accessible in Sect. 2. In Sect. 3, we have illustrated system model and design goals for our proposed methodology. In the next section, we have presented our novel disaster management system. Demanding performance analysis of the proposed scheme is presented in Sect. 5. We conclude our work with final comments in the Sect. 6.

2 Related Works

In this section, we will provide an analysis of several existing proposals related to this field. Several traffic management schemes are already there at work using different network communications. Several researchers have proposed different specific architectures, data collection schemes, and routing algorithms for traffic management.

The proposed approach is to accumulate high-quality travel time using dynamic vehicle traffic monitoring [2]. DTMon uses task organizers to interconnect with moving vehicles. Virtual strips are used as the collection points of traffic data on roads. The vehicle is equipped with a communication module that communicates with the VS and the traffic monitoring control. They examined the capability of the DTMon using the VANET modules. By assigning various tasks and with multiple locations, the message reception ratio and information receiving ratio are evaluated. The results of the simulation show better DTMon performance than an automatic vehicle location system in terms of monitoring. The use of virtual strips in DTMon can be extended for the detection and tracking of the tail end, caused by congestion.

Bruno and Nurchis [3] proposed two efficient algorithms for data collection: GREEDY and Probabilistic Data Collection for multimedia network vehicle sensors. They simulated the proposed algorithm using a VANET MobiSim and NS2 simulator. The results of the simulation show that the GREEDY solution can achieve a more uniform coverage and consume less network bandwidth.

Chao and Chen [4] proposed an intelligent traffic management system based on RFID for seminal traffic flow. The proposed scheme uses an RFID system that complies with the IEEE 802.11p protocol to detect the number of vehicles and calculate the time. The scheme used the Zigbee modules to send real-time data, like weather conditions and vehicle registration information, to the regional control center. The proposed system can perform remote transmission and reduce traffic accidents.

Friesen et al. [5] developed a complete data collection system that employs a variety of wireless networking technologies and devices to collect inferred traffic data at an intersection along a main street in an urban setting. The vehicle's presence and vehicle's path information is collected by a discovery of the Bluetooth device, and the information is transmitted to the master node through the IEEE 802.15.4 protocol. To obtain real-time information about the intersection, the master node sends the data to the server every specified time interval.

Du et al. [6] proposed a system for monitoring and estimating traffic for Shanghai using VSN. They have proposed a patrol circuit and control algorithms greedy patrol to improve performance completion matrix based traffic monitoring. The simulation results have shown that the proposed algorithms reduce the traffic estimation error from 35 to 10% compared to the random patrol method.

The information communicated by vehicles should be secured. Many researchers have already published many research papers addressing the issue of security of vehicular ad hoc networks. In [7], Kalid Elmufti et al. proposed a time-based authentication mechanism using VANET. Mona Kamel et al. proposed a secure remote patient surveillance system in [8]. The authors have used mobile phones as the core of their monitoring system 'anywhere, anytime.' Using Bluetooth and GSM/GPRS technology, authors have designed a low-cost, secure system for vital data acquisition and visualization in mobile devices.

In [9], Dusit Niyato et al. presented an architecture for the remote patient monitoring service based on heterogeneous wireless access. Barua et al. [10] proposed a secure and QoS WBAN assurance system in the application of e-Health, which is a widely used technique in remote patient control system. In [11] Qing Ding et al. proposed an event-based reputation model for filtering malicious messages. In [11], a dynamic role-dependent reputation assessment mechanism is presented to determine whether an incoming traffic message is meaningful and trustworthy to the driver. In [11], all vehicles find the same traffic event in different functions. In

[12], M. Masi et al. proposed protocol based on integrating the healthcare enterprise specification for the authentication of health professionals and the provision of rural health care.

In [13], X. Liang et al. have designed a distributed prediction-based secure and reliable routing protocol which is wireless body area network-based solution for remote health care. This protocol can be used to improve reliability and prevent data-injection attacks during data communication. SAGE [14] is a privacy preservation scheme proposed by X. Lin. et al., which can achieve not only content oriented to privacy, but also contextual privacy against a strong global adversary. In [15], A. G. Logan et al. proposed a remote patient monitoring system especially for measuring hypertension in diabetic patients. In [16], G. Yan et al. have proposed a probabilistic routing algorithm for VANET, which is based on the probability model for the duration of the link based on realistic vehicle dynamics and radio-propagation assumptions. FSR [17] is a proactive or table-based routing protocol, which relies on link state routing and improved global state routing. This protocol is very poor in small ad hoc networks and has much less knowledge about distant nodes.

VAAD [18] is developed on the framework and forward methodology by using predictable vehicle mobility and suitable for multi-hop data delivery relationship. In the scenario of topology change and traffic density, there is a long delay. GeOpps [19] is another approach in which the navigation system suggested vehicle routes to select the next hop node, and the navigation information for a node is revealed to the network, which may affect the privacy of the node concerned. GRANT [20] applies the extended greedy routing technique to avoid local maxima and divides the abstract neighborhood of the table plane into the areas and includes per area only one representative neighbor.

Connectivity-aware routing [21] has been designed for the city and/or road environment and uses AODV [22] for path discovery and uses the group's preferred broadcast mode for broadcast mode data. This protocol has overcome the problem of local maxima and can ensure shortest distance from the source to the destination path connected. Chen et al. [23] developed a routing protocol based on diagonal intersection to overcome the problem of CAR and builds a series of diagonal intersections between vehicles' source and destination. Taleb et al. [24] proposed ROMSGP to improve the routing protocol in an urban environment that has been identified that unstable routing would normally occur due to loss of connectivity if a vehicle moves outside the transmission range of a neighboring vehicle.

In the above, we have discussed about some existing research works that have been implemented to ensure the control of road traffic. We have also studied some existing traffic management mechanisms by performing congestion control. We have also considered some safety-related issues required for VANET communication in the management of emergency situations. After an exhaustive analysis of previous algorithms, we deliberately feel a need for a system that can meet the following goals:

1. Provide instant medical relief to persons suffering from accident or any other medical emergency.

2. Provide help regarding fire-related disasters.
3. Inform nearest police station in case of any civil problem occur.
4. Inform local municipality about conditions of road.

In the next section, we have discussed our proposed system model and design objectives for achieving the above-mentioned objectives.

3 Models and Design Goals

In this section, we have discussed the proposed system model and tried to identify the design requirement.

3.1 System Model

In our proposed proposal, we explore VANET application in urban scenario and we have presented the proposed disaster management system. For disaster management system, we rely on IEEE802.11p, namely VANET, for vehicle to Local Registration Authority (LRA), Event Management Center (EMC) to rescue cars (RC), and LRA to RC communications. For Global Registration Authority (GRA) to LRA, GRA to EMC, and LRA to EMC communications, we have used Wi-Fi/IEEE 802.11n. DMS consists of five interactive entities:

- Global Registration Authority (GRA) issues an initial trust value and a unique sequence number to each new vehicle. It receives all event messages and assigns a unique event identifier to each and every event. In the proposed scheme, GRA is trusted by the all participants and is in charge of the vehicles, EMCs, and rescue cars registration. It is also connected to LRAs. GRA has implicit power with ample computing and storage capabilities and infeasible for any adversary to compromise.
- Event Management Center (EMC) is in charge of rescue cars used for disaster management. EMC has used for reporting the event message and requesting the current event identifier from the GRA. Authorized service providers such as hospitals, fire stations, and police stations can act as EMCs.
- Local Registration Authority (LRA) maintains information of all vehicles registered under it and updates trust value for these vehicles after a certain interval. It is the receiver at precisely known fixed location which is used to derive correct information for nearby expedient receivers. RSUs can act as LRA. RSUs are static in nature and can be deployed where it is needed. RSU can act as a wireless access point which provides wireless accessibilities to users within the coverage area. RSUs are interconnected through Internet or by dedicated network and form a RSU backbone network. LRAs are operated under GRA and considered as trustworthy by other components. Data packets send by vehicles are securely forwarded

to corresponding EMC and GRA by using secure Internet protocol. In addition, it also checks senders' trust value. In DMS, LRAs are distributed in the urban areas where already existing network infrastructures are available.

- Rescue cars (RC) are special types of vehicles, which are registered under GRA and maintained by EMC (an ambulance, fire brigade car, etc., can act as RCs).
- For being a part of this system vehicles should be registered previously. Smart vehicles are equipped with On Board Units (OBUs). A typical OBU is equipped with a GPS along with short-range wireless communication module. It can communicate with LRA or other vehicles within the vicinity via wireless connections. Before establishing any communication among vehicular nodes, the trust value of the vehicle should be checked to avoid the communication through malicious vehicles.

3.2 Design Requirements

- **Incentive Strategy**

For designing the proposed disaster management system, participation of smart vehicle is more important. Though participation of malicious vehicles is not desirable for any system, fair participation of trusted vehicles is very much essential for successful implementation of our proposed scheme. In our proposed scheme, OBU-equipped smart vehicles are responsible for data forwarding and fair incentive scheme is very much essential for this proposal.

For the incentive calculation mechanism, we have reused our previously published trust-based strategy TruVAL [25]. In [25], trust of an agent is a discernment regarding its behavior norms, which is held by other agents, based on experiences and observation of its past actions. In the scope of investigating self-seeking behavior, the trust value of a node indicates other nodes' discernment about the cooperation of the node. Any node is considered as selfish by other nodes, i.e., the trust value of that node is low. If the trust value of a node is high, that node is identified as cooperative with other nodes. By using TruVAL [25], with monitoring selfishness of a node, we can measure the trustworthiness of each participated vehicle.

In [25], we have identified parameters depended on which trust value of any registered vehicle is updated. According to the scheme [25], the trust value can be increased when the vehicle acts as an intermediary node and successfully forwards the information toward to the destination node. In future, the trust value would help the vehicular node during the initiation of any communication.

4 Proposed Solution: Disaster Management System

The proposed scheme is illustrated as follows:

1.	Vehicle → LRA: After occurring of any event, those vehicles are present in the event spot, send an event notification to LRA
2.	LRA → GRA and LRA → EMC: LRA checks if the vehicle is trusted or not. If it is a trusted vehicle, then it forwards this message to EMC and GRA
3.	GRA → LRA and GRA → EMC: GRA generates an Event_id for the corresponding event and sends it to LRA and EMC
4.	EMC → LRA: After any event notification received by Event Management Centre (EMC), EMC communicate with Local Registration Authority (LRA) regarding the details of the event
5.	LRA → EMC: LRA sends detail information about the event.
6.	EMC → RC: From the above information sent by LRA, EMC asses the type of immediate help need at event location. EMC informs RC about the event and instructs to start rescue operation
7.	RC → LRA: RC sends a request to corresponding LRA to send the optimized trusted path to event location
8.	LRA → RC: LRA initiates a route finding procedure for finding a secure and optimized route from RC to event location. After finding the required route, LRA forwards it to RC

When any vehicle notice occurrence of any undesirable event, it use Disaster Management System (DMS) application for handle that event.

Algorithm: DMS()

Step1:	Call event_Notification (V_ID, Location_ID) function.
Step2:	Append in send_Event_queue[event_Notification_ID][TTL] = event_Notification (V_ID, Location_ID)
Step3:	LRA execute receive_notification()
Step4:	GRA execute event_Registration(V_ID,LRA_ID).
Step5:	EMC execute event_Analysis(event_id)
Step6:	RC execute event_LocationSurvey(RC_pos, event_loc)
Step7:	Set Flag = event_LocationSurvey(RC_pos, event_loc)

Step8:	If Flag == 1
	      Print " DMS Successful"
	Else
	      Print " DMS Failed"

Step9:	END

Algorithm: event_Notification (V_ID, Location_ID) /* Perform by Vehicle V /*

Step 1: Construct event_Notification message
Step 2: Send event_Notification[V_ID, Location_ID, TRlevel, Disaster_Level, LRA address, TTL] message.
Step 3: event_Notification_ID++;
Step 4: return(event_Notification_ID, TTL)
Step 5: END

Algorithm: receive_notification() /* Perform by LRA /*

Step 1: Receive event_Notification[V_ID, Location_ID, TRlevel, Disaster_Level, LRA address, TTL] message.
Step 2: set Message_Accept_Index = Trust_Check(V_ID, TRlevel)
Step 3: if(Message_Accept_Index == 0)
               Exit from the application.
   Else
          Forward event_Notification[V_ID, Location_ID, TRlevel, Disaster_Level, LRA address] message after updating TTL to GRA and EMC.
Step 4: return(0)
Step 5: END

Algorithm: Trust_Check (V_ID, TRlevel) /* Perform by LRA /*

Step 1: Send TRlevel_validate [V_ID, TRlevel] message to GRA.
Step 2: GRA send Reply[V_ID, TRlevel, Validation] message.
Step 3: if (validation = false)
                       set TRlevel --.
                       Exit from the application.
      Elseif (TRlevel < Authentication_level)
                     return(0)
      else
             return(1)
Step 4: END

Algorithm: event_Registration(V_ID,LRA_ID) /* Perform by GRA /*

Step 1: Generate event_ID
Step 2: send registration_info [event_ID, LRA_ID,V_ID] message to EMC and LRA
Step 3: End

Algorithm: event_Analysis(event_id) /* Perform by EMC /*

Step 1: Send req_detail_event[LRA address, event_id] message to LRA
Step 2: Receive reply from LRA as rep_ detail_event[LRA address, event_id,
　　　　Event_type, Event_Time, Event_Location]
Step 3: if (Event_type ==00)
　　　　　　　　Send request to RC00
　　　　if (Event_type ==01)
　　　　　　　　Send request to RC01
　　　　if (Event_type ==10)
　　　　　　　　Send request to RC10
　　　　if (Event_type ==11)
　　　　　　　　Send request to RC11
Step 4: End

See Tables 1 and 2.

Algorithm: event_Location_Survey(RC_pos, event_loc) /* Perform by RC (This is a general activity for all types of RC /*

Step1: send optimal_route_request [RC_geographic_location,
Event_geographic_location] message to LRA.
Step 2: receive optimal route from LRA
Step 3: Reach event location by following the optimal route.
Step 4: Set help_Index =1
Step 5: return(help_Index)
Step 6: END

By using the above-mentioned steps, vehicular nodes can communicate with each other through VANET (Fig. 1).

Table 1 Requirement classification table

Event_type	Requirement classification
00	Required medical help
01	Required fire brigade's help
10	Required civil-related help
11	Required road-maintenance-related help

Table 2 RC-type description table

Rescue car (RC) type	Description
RC00	Ambulance
RC01	Fire brigade
RC10	Police van
RC11	Municipality

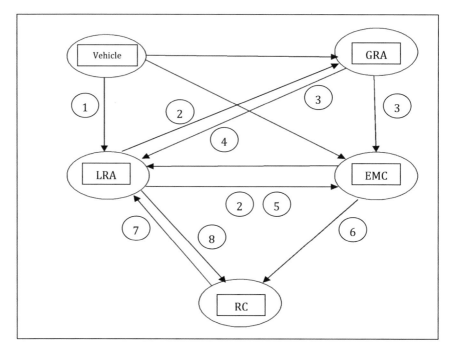

Fig. 1 Pictorial representation of disaster management system (DMS)

5 Performance Analysis

In this section, we analyze the performance of our proposed scheme using a probabilistic module and simulation has performed using NS2.

5.1 Probabilistic Analysis

The success rate of the proposed scheme is very much depended on the participation of smart vehicle. For measuring effectiveness of our proposed algorithm, we have conducted a probabilistic analysis for finding the relation among the total number of users (vehicles) in a specified region, probability (P_c) of agreed vehicle to participate in message communication, and probability (P_m) of successful message forwarding.

Let n is the total number of vehicle present in the region. $E(A_c)$ is an event, where c is the number of cooperative vehicle and $n - c$ be the number of non-cooperative vehicle. Let $E(P_m)$ be the event that there is at least one vehicle that agrees to communicate with its immediate LRA. Using the equation, the relation among $Pr(P_m)$, n, and $Pr(P_c)$ has established to measure the probability of vehicles' participation in the system.

Table 3 Simulation environment parameters

Parameter	Value
Area	10000 m × 6000 m
Number of nodes	50 (min), 100 (max)
Velocity	40 km/h(min), 80 km/h(max)
Packet interval	Every 20 min
RSUs	10 km^2
OBUs	250 m
Simulation time	12 h
Channel type	Wireless channel
Radio-propagation model	Two-ray ground
Antenna model	Omni antenna
Network interface type	Wireless phy
Mac type	802.11

$$Pr(P_m) = \sum_{c=0}^{n} Pr(E(P_m)|E(A_c)) \cdot Pr(E(A_c))$$
$$= 1 + (1 - P_c)^n - 2(1 - P_c/2)^n$$

Here, $(1 - P_c)^c$ is the probability that none of the vehicle communicates with LRA about the event. $(1 - (1 - P_c)^c)$ is the probability that there is at least one cooperative node, and $(1 - (1 - P_c)^{n-c})$ is the probability that there will be at least one non-cooperative node. Hence, $Pr(E(A_c)) = \left(\frac{1}{2}\right)^c \left(1 - \frac{1}{2}\right)$, and $Pr(E(Pm)|E(Ac)) = (1 - (1 - P_c)^c) \cdot (1 - (1 - P_c)^{n-c})$ each user position is independent.

5.2 Simulation

The proposed algorithm is implemented using NS2 simulator. The simulation model consists of a network model that has a number of wireless nodes, which represents the entire network to be simulated (Table 3).

We have chosen performance metrics with respect to our basic objectives. Measurement of QoS in this system can be done by achieving a cumulative measure of reliability, energy efficiency, timeliness, robustness, availability, and security. For measuring the above-mentioned performance indices, the followings are particularly significant.

End-to-end delay: The average time taken by a data packet to reach in the destination and also includes the delay caused by route discovery process (Processing delay), the queue in data packet transmission (Queuing delay), the delay required to push all the bits in a packet on the transmission medium in use (Transmission delay) and the delay required for the bit to propagate to the end of its physical trajectory

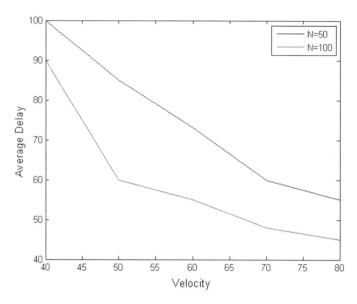

Fig. 2 Average end-to-end delay

(Propagation delay). Data packets only those successfully delivered to destinations are counted.

Processing delay (ρ) can be measured as follows: $\rho = (H + (D/P)) /C$, where C is the capacity of the links in bps, D is the bit size of application data, H is the header size in bits, and P is the number of packets for this application data.

For measuring queuing delay (λ), we have used Little's theorem. Little's theorem provides a relation between the average number of packet in the system (N), the arrival rate (T), and the delay (λ), given by $N = \lambda T$, $\rightarrow \lambda = N/T$.

Transmission delay (τ) can be formulated as follows: $\tau = S/N$, where N = number of bits and S = size of packet.

Propagation delay (PD) can be calculated using the formula given below. PD = d/s, where d = distance and s = propagation velocity.

The propagation velocity depends mainly on the physical distance between source and destination. Mostly it is close to the speed of light.

End-to-end delay is the summation of all the above-mentioned delay, i.e., end-to-end delay = $\rho + \lambda + \tau + PD$.

Figure 2 depicts the relationship between average end-to-end delay and velocity having variation in number of nodes (N). We have observed that by increasing N, we can reduce average delay. The average delay varies between 45 and 90 min having N = 100, and it varies from 55 to 100 min having N = 50.

Packet delivery ratio is the percentage of data packets delivered from source to destination and closely related to the reliability of the network. It can be used to represent the probability of packets being received. A packet may be lost due to several constraints, such as congestion, bit error, or bad connectivity.

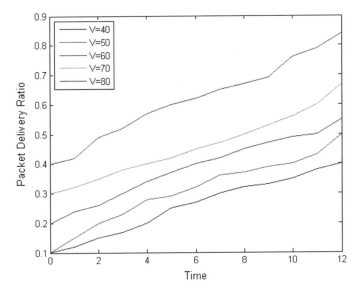

Fig. 3 Packet delivery ratio

Figure 3 shows the impact of vehicle speed on the packet delivery ratio. According to Fig. 3, we can observe that better packet delivery ratio would be achieved by increasing vehicle's speed as well.

For example, around 40% more packet delivery ratio can be achieved by increasing vehicle's speed from 40 to 80 km/h at the second hour of simulation with 50 nodes. To ensure higher packet delivery ratio, we have to ensure higher velocity of vehicles. If number of vehicles would increase, corresponding message communication would also be increased and our proposed algorithm provides better results.

6 Conclusions

In this paper, we have proposed a disaster management system using VANET, where we need to use dynamic communication with the help of vehicular network. This proposed solution can be used by on-road vehicles on demand basis. DMS would help the vehicles who are in any undesired situation. The proposed system is simulated using NS2 simulator. The simulation result shows our algorithm works better in a high-speed environment. The proposed robust solution performed better in urban scenarios.

References

1. Abdelhamid, S., Hassanein, H.S., Takahara, G.: Vehicle as a Resource (VaaR). IEEE Netw. **29**(1), 12–17 (2015)
2. Arbabi, H., Weigle, C.M.: Using DTMon to monitor transient flow traffic. In: Proceedings of the IEEE Vehicular Networking Conference (VNC), Jersey City, NJ, USA, 13–15 Dec 2010, pp. 110–117
3. Bruno, R., Nurchis, M.: Robust and efficient data collection schemes for vehicular multimedia sensor networks. In: Proceedings of the IEEE 14th International Symposium and Workshops on World of Wireless, Mobile and Multimedia Networks (WoWMoM), Madrid, Spain, 4–7 June 2013 pp. 1–10
4. Chao, K.H., Chen, P.: An intelligent traffic flow control system based on radio frequency identification and wireless sensor networks. Int. J. Distrib. Sens. Netw. 1–10 (2014)
5. Friesen, M., Jacob, R., Grestoni, P., Mailey, T., Friesen, R.M., McLeod, D.R.: Vehicular traffic monitoring using Bluetooth scanning over a wireless sensor networks. Can. J. Electr. Comput. Eng. **37**, 135–144 (2014)
6. Du, R., Chen, C., Yang, B., Lu, N., Guan, X., Shen, X.: Effective urban traffic monitoring by vehicular sensor networks. IEEE Trans. Veh. Technol. **64**, 273–286 (2015)
7. Elmufti, K., Weerasinghe, D., Rajarajan, M., Rakocevic, V., Khan, S.: Timestamp authentication protocol for remote monitoring in eHealth. In: 2nd International conference on Pervasive Healthcare, pp. 73–76. IEEE, Tampare, Finland (2008)
8. Kamel, M., Fawzy, S., El-Bialy, A., Kandil, A.: Secure remote patient monitoring system. In: 1st Middle East Conference on Biomedical Engineering (MECBME), pp. 339–342. IEEE, Sharjah, UAE (2011)
9. Niyato, D., Hossain, E., Camorlinga, S.: Remote patient monitoring service using heterogeneous wireless access networks: architecture and optimization. IEEE J. Sel. Areas Commun. **27**(4), 412–423 (2009)
10. Barua, M., Alam, M.S., Liang, X., Shen, X.: Secure and quality of service assurance scheduling scheme for WBAN with application to e-Health. In: Wireless Communications and Networking Conference (WCNC), pp. 1102–1106. IEEE, Cancun, Quintana Roo (2011)
11. Ding, Q., Li, X., Jiang, M., Zhou, X.: Reputation-based trust model in vehicular Ad Hoc networks. In: Wireless Communications and Signal Processing (WCSP), pp. 1–6. IEEE, Suzhou, China (2010)
12. Masi, M., Pugliese, R., Tiezzi, F.: A standard-driven communication protocol for disconnected clinics in rural areas. In: 13th International Conference on e-Health Networking, Applications and Services, pp. 304–311. IEEE, Columbia, MO (2011)
13. Liang, X., Li, X., Shen, Q., Lu, R., Lin, X.: Exploiting prediction to enable secure and reliable routing in wireless body area networks. In: The 31st International Conference on Computer (2012)
14. Lin, X., Lu, R., Shen, X., Nemoto, Y., Kato, N.: SAGE: a strong privacy-preserving scheme against global eavesdropping for ehealth systems. IEEE J. Sel. Areas Commun. **27**(4), 365–378 (2009)
15. Logan, A.G., McIsaac, W.J., Tisler, A., Irvine, M.J., Saunders, A., Dunai, A., Rizo, C.A., Feig, D.S., Hamill, M., Trudel, M., Cafazzo, J.A.: Mobile phone-based remote patient monitoring system for management of hypertension in diabetic patients. Am. J. Hypertens. **20**(9), 942–948 (2007)
16. Yan, G., Olariu, S., Salleh, S.: A probabilistic routing protocol for VANET. Int. J. Mob. Comput. Multimed. Commun. (IJMCMC) **2**(4), 21–37 (2010)
17. Pei, G., Gerla, M., Chen, T.W.: Fisheye state routing: a routing scheme for ad hoc wireless networks. In: International Conference on Communications, New Orleans, LA, vol. 1, pp. 70–74 (2000)
18. Zhao, J., Cao, G.: VADD: vehicle-assisted data delivery in vehicular ad hoc networks. IEEE Trans. Veh. Technol. **57**(3), 1910–1922 (2008)

19. Leontiadis, I., Mascolo, C.: GeOpps: geographical opportunistic routing for vehicular networks. In: IEEE International Symposium on World of Wireless, Mobile and Multimedia Networks (WoWMoM), Espoo, Finland, pp. 1–6 (2007)
20. Arbabi, H., Weigle, C.M.: Monitoring free flow traffic using vehicular networks. In: Proceedings of the IEEE Consumer Communications and Networking Conference (CCNC), Las Vegas, NV, USA, 9–12 Jan 2011, pp. 272–276
21. Schnaufer, S., Effelsberg, W.: Position-based unicast routing for city scenarios. In: IEEE International symposium on World of Wireless, Mobile and Multimedia Networks (WoWMoM), Newport Beach, CA, pp. 1–8 (2008)
22. Perkins, C., Belding-Royer, E., Das, S.: Ad hoc on-demand distance vector (AODV) routing. In: 2nd IEEE Workshop on Mobile Computing Systems and Applications, WMCSA, New Orleans, LA, pp. 90–100 (1999)
23. Chen, Y.S., Lin, Y.W., Pan, C.Y.: A diagonal-intersection-based routing protocol for urban vehicular ad hoc networks. Telecommun. Syst. **46**(4), 299–316 (2010)
24. Taleb, T., Sakhaee, E., Jamalipour, A., Hashimoto, K., Kato, N., Nemoto, Y.: A stable routing protocol to support ITS services in VANET networks. IEEE Trans. Veh. Technol. **56**(6), 3337–33347 (2007)
25. DasGupta, S., Chaki, R., Choudhury, S.: TruVAL: trusted vehicle authentication logic for VANET. In: 3rd International Conference on Advances in Computing, Communication and Control (ICAC3), vol. 361, pp. 309–322. Springer Link, Mumbai, India (2013)

Part III
Pattern Recognition

A Computer-Aided Hybrid Framework for Early Diagnosis of Breast Cancer

Sourav Pramanik, Debapriya Banik, Debotosh Bhattacharjee
and Mita Nasipuri

Abstract We have presented here a novel framework for the early diagnosis of breast cancer. The framework is comprised of two major phases. Firstly, the potential suspicious regions are automatically segmented from the breast thermograms. In the second phase, the segmented suspicious regions are diagnostically classified into benign and malignant cases. For the automatic segmentation of the suspicious regions, a new region-based level-set method named GRL-LSM has been proposed. Initially, the potential suspicious regions are estimated by the proposed adaptive thresholding method (ATM), named GRL. Then, a region-based level set method (LSM) is employed to precisely segment the potentially suspicious regions. As initialization plays a vital role in a region-based LSM, so we have proposed a new automatic initialization technique based on the outcome of our adaptive thresholding method. Moreover, a stopping criterion is proposed to stop the LSM. After the segmentation phase, some higher-order statistical and GLCM-based texture features are extracted and fed into a three-layered feed-forward artificial neural network for classifying the breast thermograms. Fifty breast thermograms with confirmed hot spots are randomly chosen from the DMR-IR database for the experimental purpose. Experimental evaluation shows that our proposed framework can differentiate between malignant and benign breasts with an accuracy of 89.4%, the sensitivity of 86%, and specificity of 90%. Additionally, our segmentation results are validated quantitatively and qualitatively with the respective breast thermograms which were manually delineated by two experts and also with some classical segmentation methods.

S. Pramanik (✉) · D. Banik · D. Bhattacharjee · M. Nasipuri
Department of Computer Science and Engineering, Jadavpur University, Kolkata 700032, India
e-mail: srv.pramanik03327@gmail.com

D. Banik
e-mail: debu.cse88@gmail.com

D. Bhattacharjee
e-mail: debotosh@cse.jdvu.ac.in

M. Nasipuri
e-mail: mnasipuri@cse.jdvu.ac.in

© Springer Nature Singapore Pte Ltd. 2019
R. Chaki et al. (eds.), *Advanced Computing and Systems for Security*,
Advances in Intelligent Systems and Computing 883,
https://doi.org/10.1007/978-981-13-3702-4_7

Keywords Suspicious region segmentation · GRL-LSM · Feature extraction
FANN · Malignant and benign breast thermogram

1 Introduction

Globally, breast cancer is the most commonly diagnosed cancer in women after
lung cancer. More than a million of women worldwide are identified with breast
cancer every year, which accounts more than 25% of all cases of female cancer
[1]. Due to this exponential growth, there is a huge demand for the development
of new technologies for the prevention of breast cancer in its early stages. Medical
infrared breast thermography has emerged as a promising tool in early diagnosis of
breast cancer patients [2]. For the last few decades, there were different researchers
in the domain of analysis of breast thermograms to differentiate between normal and
abnormal breasts. However, it is very rare to find any allied research to differentiate
between malignant and benign breasts. To analyze the acquired breast thermograms,
it is very crucial to segment the breast region from the whole image [3]. Due to
the vague nature of breast thermograms; most research groups prefer manual or
semi-automatic segmentation rather than automatic segmentation [4]. However, the
automatic segmentation technique is very much important for the computer-aided
analysis of the breast thermograms.

Sathish et al. [4] presented a new segmentation method based on canny edge
detection technique to segment the left and the right breast thermogram. Then, for
analysis, some features were extracted and fed to a classifier to differentiate between
the normal and the abnormal breast thermograms. Prabha et al. [5] made an attempt
for the segmentation of the breast thermograms by the reaction-diffusion-based level-
set method. After segmentation, authors have extracted some structural-based texture
features to perform asymmetry analysis. Mejia et al. [6] proposed a segmentation
technique by thresholding and morphological operations. Furthermore, they have
also extracted some basic textural features and differentiated the normal breast thermograms from the abnormal ones.

Apart from the breast region segmentation techniques as discussed above, some
studies have shown that the suspicious regions (hot regions) captured by an infrared
(IR) camera can be a much more significant indicator to justify the level of abnor-
malities in the breast [7]. Computer-assisted analysis of the suspicious regions after
delineating it from the whole breast area defines the degree of malignancy and the
level of expansion of the tumor [8]. However, segmentation of the suspicious regions
from the thermal breast image is a very challenging task. Because they vary in size and
suffer from intensity variation. Moreover, poor edge definition and the chaotic structure of the area of the suspicious region make it difficult to segment. In light of such
challenges, considerable progress has been made to segment the suspicious regions,
but the results have not been reliable so far. Etehad Tavakol et al. [9] have used k-

means and fuzzy c-means for suspicious regions segmentation. In [10], authors have reported a technique for suspicious region segmentation using minimum variance quantization method followed by morphological operations. However, they did not mention the number of iteration, quantization levels, and radius of the structuring element. In [11], authors have compared three different methods, such as k-means, fuzzy c-means, and level set for the segmentation of suspicious regions and have shown that the level-set method works considerably well in comparison with the other methods. However, in this work, authors have not mentioned how they have initialized the contour points for the LSM. In [12], authors have converted the color breast thermal images to CIELUV color space, then fuzzy c-means clustering technique is used to segment different color regions. Although abnormality cases were classified the major drawback was that malignant cases could not be differentiable from benign cases by their method.

As conferred above, it is very rare to find any research related to the analysis of breast thermograms to differentiate between malignant and benign cases. So, in this study, we have proposed a framework to differentiate between malignant and benign breast thermograms. The framework basically includes two major phases. In the first phase, the potentially suspicious regions are automatically segmented from the breast thermograms. To accomplish this task, a new method GRL-LSM has been proposed. The proposed method comprises of two steps. Initially, the estimation of the potentially suspicious regions is performed by an adaptive thresholding method (ATM) based on global and local image information (GRL). Then, a region-based LSM [13] is employed to accurately segment the potentially suspicious regions. As breast thermograms often suffer from intensity-inhomogeneity and complex background, initialization of a region-based LSM based on global region information often fails to give satisfiable results. Hence, we have proposed a new automatic initialization technique based on the outcome of our initial adaptive thresholding method. In the second phase of our framework, the segmented breast thermograms are analyzed and classified as benign or malignant. To do so, some higher-order statistical features and GLCM-based texture features are extracted from the segmented left and right breasts and fed into a three-layered feed-forward artificial neural network (FANN) with gradient descent training rule. The experiments were carried out on DMR-IR database by randomly choosing 50 breast thermograms (33 benign and 17 malignant) with confirmed hot spots. Our experimental evaluation confirms that our proposed framework can differentiate between the malignant and benign breast more precisely. Additionally, we have validated our segmentation results both quantitatively and qualitatively with the respective breast thermograms which were manually delineated by two medical experts and also with the two classical segmentation methods [9]. It can be observed that our results hold good agreement with the ground truths irrespective of the irregular shape/weak edges of the suspicious regions.

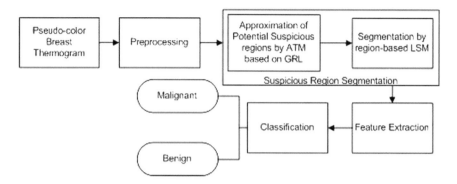

Fig. 1 Flow diagram of the proposed framework

The rest of the paper is arranged as follows. Section 2 demonstrates the proposed system in detail. Section 3 describes the details of an experiment conducted to justify the performance of the proposed framework, and the conclusion is drawn in Sect. 4.

2 Proposed System

Figure 1 shows the flow diagram of the proposed breast abnormality prediction system. Each of the steps will be described in the following subsections.

2.1 Preprocessing

Usually, in the original pseudo-color breast thermogram, colorscale also appears alongside the image, as shown in Fig. 2a, which may mislead the suspicious regions segmentation process. Thus, at the very first step, we manually remove the colorscale from the original breast thermal image, as shown in Fig. 2b. Then, the color breast thermal image is transformed into blue-channel color space as shown in Fig. 2c. From Fig. 2c, it can be seen that the transformation process preserves the location and shape of the thermal features and objects as it is a point-wise transformation. Moreover, most importantly, it has been experimentally seen that the blue-channel image in comparison with the other channels increases the contrast of the high-temperature regions and decreases the contrast of the low-temperature regions.

(a)   **(b)**   **(c)**

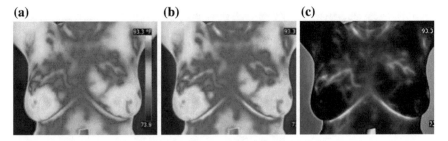

Fig. 2 **a** Original pseudo-color breast thermogram, **b** colorscale-removed image, and **c** corresponding blue-channel image

2.2 Suspicious Regions Segmentation

2.2.1 Approximation of Potentially Suspicious Regions

This method is basically accomplished in two steps: It starts with an approximation of the suspicious regions (hot regions) by an ATM and followed by isolation of the potentially suspicious regions analogous to the suspicious regions. For the approximation of the hot regions, we have proposed here a new ATM based on global-image information regularized by local-image information (GRL). Let us assume that $I^{(b)} : \omega \rightarrow \mathcal{R}^+$ be the blue-channel thermal breast image, where $\omega \subset \mathcal{R}^2$ is the image domain. In this proposed method, a threshold value is adaptively calculated for each pixel in the blue-channel image $I^{(b)}$ to identify it as a suspicious region pixel or a normal region pixel. Usually, the intensities in the center area of a suspicious region have superiority over the entire image and which gradually decreases toward the boundary. Thus, the pixel belonging to the center area of the suspicious region can be easily classified as a suspicious region pixel using a global threshold value. However, the global threshold will not work when the pixel in the suspicious region has a very small difference in intensity with the normal tissue pixel. In this case, the classification of such pixel completely depends on its neighboring pixels. Thus, we have used a combination of the global superiority and the local dependency of a pixel in the image to compute a threshold for it. The threshold value T(x, y) for the pixel $p_c \in I^{(b)}$ is computed by Eq. (1).

$$T(x, y) = I_{avg}^{(b)} + \delta * \left(I_{max}^{(b)} - I_{min}^{(b)} \right) \tag{1}$$

$$\delta = \frac{1}{|w|} \sum\nolimits_{p_i \in w(p_c)} s(p_c == p_i) \tag{2}$$

$$s(\cdot) = \begin{cases} 1, & \text{if } p_c == p_i \\ 0, & \text{otherwise} \end{cases} \tag{3}$$

Fig. 3 Different probably
high-intensity zones in a
breast thermogram

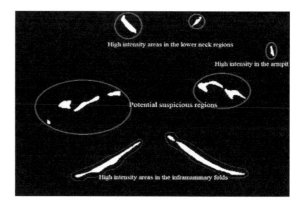

where, $I_{avg}^{(b)}$ is an average intensity of the image $I^{(b)}$, $I_{max}^{(b)}$ and $I_{min}^{(b)}$ are the maximum
and minimum intensity of the image $I^{(b)}$, δ is a thresholding control parameter, $|w|$
is the size of the square window around the centered pixel p_c, and (x, y) is the
coordinate of the pixel p_c. In our case, a 3×3 window size is considered. Now,
if a pixel $p_c(x, y) \in I^{(b)} \geq T(x, y)$, it is classified as the suspicious region pixel;
otherwise, it is classified as a normal tissue region pixel.

Once the suspicious regions (or hot regions) are approximated, the next goal is to
identify the potentially suspicious regions. Let B_{SR} be the binary image containing
the hot region areas, as shown in Fig. 3. In Fig. 3, a white region encircled with red
colors exhibits the hot region areas and the possible suspicious regions are located
approximately at the center part of the image. Therefore, if we calculate the centroids
of all the hot regions, the centroids of the possible suspicious regions will be greater
than the centroids of the regions located in the lower-neck areas and armpits. Similarly, it is less than the centroid of the inframammary fold regions. Hence, in this
paper, potentially suspicious regions are isolated based on the centroid information.
Note that the proposed adaptive thresholding method based on GRL approximates
the highest intensity zones in the potentially suspicious regions. Therefore, to accurately segment the potentially suspicious regions, GRL results are used to initialize
the LSM which we have discussed in the next section.

2.2.2 Segmentation by Region-Based Level-Set Method (LSM)

To precisely segment the potentially suspicious regions from the breast thermograms,
we have used the Chan-Vese (CV) level-set method (LSM) [13]. Basically, the CV-
LSM is a region-based method where contours are driven on the basis of intensity,
color or texture information. The suspicious regions in a breast thermogram are often
susceptible to noise, poor edge definition, and have irregular shapes. So to address
the aforesaid problems, CV-LSM is employed as this method gives robust results

despite such issues [14]. In the CV-level-set method, the energy functional is defined as follows:

$$E_{CV}(\varphi, c_i) = \beta \int_\omega \delta_\varepsilon(\varphi(x, y)) |\nabla\varphi(x, y)| dxdy$$

$$+ \sum_{i=1}^{2} \alpha_i \int_\omega \left| I^B(x, y) - c_i \right|^2 H^i dxdy \tag{4}$$

where $\varphi : \omega \rightarrow \mathcal{R}$ is the level-set function which is used to implicitly represent the closed contour $C(q) : \mathcal{R} \rightarrow \omega$, c_1 and c_1 are two constants computed inside and outside of C as the average intensity and play a very crucial role for the evolution of the contour C, α_1 and α_2 are two positive-fixed parameters used as trade-off between the first and second terms of the data fitting term in (4), $H^1 = H_\varepsilon(\varphi(x, y))$ and $H^2 = (1 - H_\varepsilon(\varphi(x, y)))$ are the smooth approximation of the Heaviside function, the first term on the right-hand side of (4) is called the contour length term which is used to smooth the φ and the second term is called data fitting term that is used to drive the φ toward the object boundary, δ_ε signifies the Dirac delta function. Typically, in this work, let us assume that the zero-level set φ is negative inside the closed contour C, positive outside of C, and zero on the C, respectively [15].

The initialization of the level-set function plays a crucial role in driving the level set for accurate detection of the suspicious regions. In practice, if the initialization is apart from the original region of interest boundaries, the zero level set will take huge iterations to segment the region of interest. Most of the works used manual initialization, which required thorough domain knowledge and also a very much labor-intensive procedure. Therefore, in this work, the outcome of the ATM is used here to precisely initialize the level-set function (LSF) φ. The LSF is initialized as follows: Let, B represents the binary image containing different potentially suspicious regions as obtained from the previous section.

To initialize the level-set function $\varphi(x, y)$, B should satisfy the property of φ [15], so we have computed Euclidean distance transform (EDT) on the binary image B(x, y) as in Eq. (6)

Let $p(x, y) \in B$ be any pixel in B and $q_n(x, y) \in B$ be non-zero pixels in $\{B|q_n(x, y) = 1\}$

$$B_{EDT}^p(x, y) = \underset{n=1,2,...,L}{Min} \|p - q_n\| \tag{6}$$

where L is the number of non-zero pixels in B and ‖.‖ denotes the Euclidean distance.

For a binary image B(x, y) in Fig. 4, where the zero pixels represent the background and the non-zero pixels enclosed with red color represent a potentially suspicious region. For each pixel in B(x, y) EDT is computed and mapped to $B_{EDT}(x, y)$. Finally, $B_{EDT}(x, y)$ is initialized to the level-set function to drive the level set.

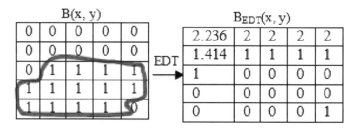

Fig. 4 Euclidean distance transform of B(x, y)

Now, the initial zero-level-set function φ can be defined as follows:

$$\varphi(x, y, t = 0) = B_{EDT}^p(x, y) \tag{7}$$

In each iteration, the energy function (4) is minimized with respect to φ and c_i by considering the Euler–Lagrange equations and updating φ using the gradient descent method:

$$\frac{\partial \varphi(x, y)}{\partial t} = \delta_\varepsilon(\varphi(x, y)) \left[-\alpha_1 (I^B(x, y) - c_1)^2 + \alpha_2 (I^B(x, y) - c_2)^2 + \beta div \left(\frac{\nabla \varphi}{|\nabla \varphi|} \right) \right]$$

$$\varphi(x, y, t = 0) = B_{EDT}^p(x, y) \tag{8}$$

Similarly, c_1 and c_2 are updated as follows:

$$c_i = \frac{\int_\omega I^{(b)}(x, y) H^i dx dy}{\int_\omega H^i dx dy}$$

Generally, in region-based CV-LSM method, the criteria to stop the curve evolution are set manually by stating the value of N (no. of iterations) or by minimizing the energy term for the level-set function. The drawback of the prior stopping terms in the domain of thermal breast imaging is that the image is prone to intensity variation and weak edges, which may lead to over or under segmentation. So, we have proposed a new technique to stop the curve automatically.

Let, the level-set function for $(N - 1)$th, Nth, and $(N + 1)$th iteration be $\varphi_{(x,y)}^{N-1}$, $\varphi_{(x,y)}^N$ and $\varphi_{(x,y)}^{N+1}$. The curve will stop automatically if the correlation coefficient (r) satisfies for $N > 1$ Eq. (9)

$$r(\varphi_{(x,y)}^{N-1}, \varphi_{(x,y)}^N) \approx 1 \approx r(\varphi_{(x,y)}^N, \varphi_{(x,y)}^{N+1}) \tag{9}$$

Finally, the resultant curve will segment all the potentially suspicious regions more precisely. Figure 5 represents automatic segmentation result of a breast thermogram by our proposed method.

(a)   **(b)**   **(c)**

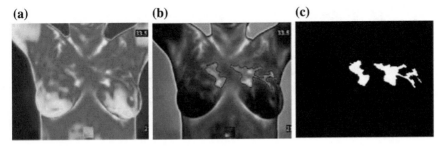

Fig. 5 **a** Initial breast thermogram, **b** segmentation by proposed method, and **c** final-segmented binary image

2.3 Feature Extraction

After the segmentation of the potentially suspicious regions, a set of features is extracted from each of the breast (left and right) of each patient. Here, we have considered two variants of feature sets: GLCM-based texture features and higher-order statistical moments. GLCM of an image is the measure of how often various combinations of gray levels co-occur in an image. From the GLCM, we have computed entropy, contrast, correlation, energy, and homogeneity [4]. It may be noted that texture analysis plays a crucial role in the areas of infrared image analysis since temperature variations in thermogram images are represented by its texture. Apart from the textural features, the statistical moments are also calculated up to order 20. The first-order moment is the mean of the pixels values in the block and is calculated using Eq. (10) [16]. The remaining higher-order normalized moments are calculated by Eq. (11)

$$M^{(1)} = \frac{1}{N} \sum_{i=1}^{N} x_i \tag{10}$$

$$M^{(q)} = \frac{1}{\sigma^2} \sum_{i=1}^{N} \left(x_i - M^{(1)} \right)^q \tag{11}$$

where N is the number of non-zero pixels in the region of interest, $q = 2, 3, \ldots, 20$ and σ is the standard deviation.

After the computation of the five GLCM-based texture features and statistical moments up to order 20, they are concatenated to form a 25-element feature vector for each left and right breast.

Let $\left[f_v^{(L)} \right]_{1 \times 25}$ and $\left[f_v^{(R)} \right]_{1 \times 25}$ demonstrate the feature vectors of left and right breast thermogram of each patient. The vectors $f_v^{(L)}$ and $f_v^{(R)}$ are defined as

$$f_v^{(L)} = \left[f_1^{L\_GLCM}, f_2^{L\_GLCM}, \ldots, f_5^{L\_GLCM}, f_1^{L\_M}, f_2^{L\_M}, \ldots, f_{20}^{L\_M} \right] \tag{12}$$

$$f_v^{(R)} = \left[f_1^{R\_GLCM}, f_2^{R\_GLCM}, \ldots, f_5^{R\_GLCM}, f_1^{R\_M}, f_2^{R\_M}, \ldots, f_{20}^{R\_M} \right] \qquad (13)$$

Finally, the asymmetry features are calculated for each patient's breast thermogram by taking the absolute difference between $f_v^{(L)}$ and $f_v^{(R)}$. Let $[F]_{1 \times 25}$ be the asymmetry feature vector of a patient's breast thermogram and is defined as

$$F = \left| f_v^{(L)} - f_v^{(R)} \right| \qquad (14)$$

2.4 Classifier Design

A three-layered feed-forward artificial neural network (FANN) [17] is used to classify the breast thermograms into malignant and benign. A total of 25 neurons in the input layer of the network are considered to fit the 25-element feature vector. The linear transformation function is used in the input layer. Here, one hidden layer is considered that consists of 50 neurons and the sigmoid transfer function is used for all the nodes present in both the hidden and output layers. Since our problem of classification is a binary problem, we have considered one neuron in the output layer. For the training of the network, the Levenberg–Marquardt back-propagation algorithm (with learning rate = 0.1) is used. A total of 30 breast thermograms (10 malignant and 20 benign) are randomly chosen from a set of 50 thermal breast images (17 malignant and 33 benign) to train the network. Remaining breast thermograms are used to test the network.

3 Experimental Results and Discussion

3.1 Database Collection

The breast thermograms used in this research work are collected from the existing DMR-IR (Database for Mastology Research) Database [18]. DMR-IR is an open-access online database that consists of breast thermograms of 287 patients. The dataset includes breast thermograms of normal, benign, and malignant patients. From the set of benign and malignant breast thermograms, 50 breast thermograms with confirmed hot spots are randomly selected for our work of which 33 thermograms are of benign breasts and 17 thermograms are of malignant breasts.

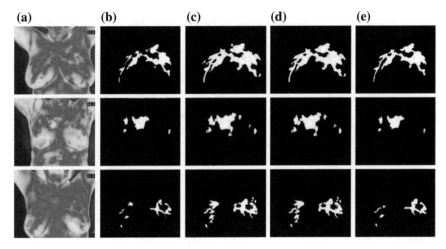

Fig. 6 **a** Breast thermal image, **b** ground truth, **c** segmentation by K-means method, **d** segmentation by Fuzzy c-means method, and **e** our proposed method

3.2 Performance Assessment

In this section, we have evaluated the effectiveness of our proposed segmentation method using qualitative and quantitative performance assessment. To compare the segmentation results, two sets of ground truth images are developed by two different medical experts in this relevant field.

For the purpose of qualitative analysis, our proposed segmentation method is compared visually with the average ground truths and also with some of the traditional segmentation methods, viz k-means and fuzzy c-means as mentioned in [9]. From Fig. 6, it can be effectually envisioned that our proposed method holds good agreement with the ground truth in comparison with the other traditional methods.

Apart from visual analysis, quantitative analysis is also accomplished to measure the accuracy of our segmentation results. Some of the traditional measures commonly used for quantitative analysis are Jaccard Index (JC), Dice Similarity (DS), Tanimoto (TN), and Volume Similarity (VS) [19]. Here we have compared our segmentation method with some of the traditional segmentation approaches as discussed above. The mean for each of the quantitative measures is evaluated for the two sets of ground truth (GT_1 and GT_2), respectively, as shown in Table 1. From Table 1, it can be seen that our proposed method noticeably outperforms when compared with the two classical segmentation methods.

Table 1 Overlap similarity measure for K-means, Fuzzy c-means, and the Proposed method

Method	JC		DS		TN		VS	
	GT_1	GT_2	GT_1	GT_2	GT_1	GT_2	GT_1	GT_2
K-means	0.571	0.572	0.716	0.717	0.925	0.926	0.779	0.776
Fuzzy c-means	0.588	0.589	0.730	0.731	0.931	0.931	0.809	0.805
Proposed method	0.699	0.704	0.820	0.823	0.964	0.965	0.932	0.935

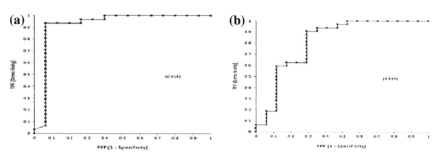

Fig. 7 a ROC curve for our method, and **b** ROC curve for a method without segmentation

3.3 Classification

In this section, we have shown the effectiveness of our segmentation results for the differentiation of the malignant and benign thermal breast images. As discussed in Sect. 2, the 25-element feature vector is formed for each patient's breast thermogram, which is then fed to the three-layered feed-forward artificial neural network (FANN) for the classification. Thus, three traditional performance measures such as accuracy, sensitivity, and specificity are calculated based on the classification results [20]. We have also quantified our classification results using receiver operating characteristic (ROC) curve. The area under the curve (AUC) is very much informative for the analysis of the system results, which typically lies between 0.5 and 1 [20]. AUC near to 1 signifies that the discriminating ability of the system is considerably well. In this work, the obtained accuracy, sensitivity, and specificity are 89.4%, 86%, and 90%, respectively. Figure 7a shows the ROC curve for the proposed method. The obtained AUC value is 0.919 and attained 100% true positive recognition at 0.4 false positive rates.

To verify the efficacy of the proposed system (suspicious regions segmentation + feature extraction) over the system without suspicious regions segmentation, the same set of features are extracted from each patient's breast thermogram without segmenting the suspicious regions and used for classification. Table 2 summarizes the performance measures of the classification results. The result confirms that our proposed method outperforms over the compared method.

Table 2 Comparison of classification performance measures

Methods	Accuracy (%)	Sensitivity (%)	Specificity (%)	Area under the ROC curve
Proposed method	89.4	86	90	0.919
Method without segmentation of the suspicious regions	78	50	93	0.814

4 Conclusion

Automatic segmentation of potentially suspicious regions from the breast thermograms plays a crucial role in the computer-assisted analysis of the breast thermograms and differentiating between benign and malignant breasts. Hence, in this paper, we have proposed a framework to automatically segment the potentially suspicious regions and analyzed the segmented regions to differentiate between benign and malignant breasts. Most of the automatic segmentation algorithm claimed till date has some human intervention either to specify the initialization point or to stop the evolution of the curve. However, our system does not require any human intervention and can segment the potentially suspicious regions precisely. The evaluation of experimental results reveals that our proposed framework can effectively differentiate between malignant and benign breasts. Moreover, our segmentation method outperforms in terms of performance measurement when compared to other traditional methods. We believe that our system can serve as a second option to a radiologist for early diagnosis of breast cancer and differentiate between the malignant and benign cases. It is significant to say that the performance of classification would have been much better if some strong feature sets were used in our study. So our future direction would be to improve the classification performance by considering powerful feature sets.

Acknowledgements Authors are thankful to DBT, Govt. of India for funding a project with Grant no. BT/533/NE/TBP/2014. Sourav Pramanik is also thankful to Ministry of Electronics and Information Technology (MeitY), Govt. of India, for providing him Ph.D.Fellowship under Visvesvaraya Ph.D. scheme.

References

1. http://www.wcrf.org/int/cancer-facts-figures/data-specific-cancers/breast-cancer-statistics. Accessed 5 Aug 2017
2. Jones, B.F.: A reappraisal of the use of infrared thermal image analysis in medicine. IEEE Trans. Med. Imaging **ED-17**, 1019–1027 (1998)
3. Francis, S.V., Sasikala, M., Saranya, S.: Detection of breast abnormality from thermo grams using curvelet transform based feature extraction. J. Med. Syst. **38**(4), 1–9 (2014)

4. Sathish, D., Kamath, S., Prasad, K., Kadavigere, R., Martis, R.J.: Asymmetry analysis of breast thermograms using automated segmentation and texture features. J. Signal Image Video Process. **10**, 1–8 (2016)

5. Prabha, S., Anandh, K.R., Sujatha, C.M., Ramakrishnan, S.: Total variation based edge enhancement for level set segmentation and asymmetry analysis in breast thermograms. In: IEEE International Conference on Engineering in Medicine and Biology Society (EMBC) (2014)

6. Mejia, T.M., Perez, M.G., Andaluz, V.H., Concil, A.: Automatic segmentation and analysis of thermograms using texture descriptors for breast cancer detection. In: IEEE International Asia-Pacific Conference on Computer Aided System Engineering (2015)

7. Etehad Tavakol, M., Ng, E.Y.K.: Breast thermography as a potential non-contact method in the early detection of cancer: a review. J. Mech. Med. Biol. **13**(2), 1–20 (2013)

8. Qi, H., Kuruganti, P.T., Snyder, W.E.: Detecting breast cancer from thermal infrared images by asymmetry analysis. In: Biomedical Engineering Handbook, pp. 27.1–27.14. CRC, Boca Raton (2016)

9. Etehad Tavakol, M., Sadri, S., Ng, E.Y.K.: Application of K- and fuzzy c-means for color segmentation of thermal infrared breast images. J. Med. Syst. **34**, 35–42 (2010)

10. Milosevic, M., Jankovic, D., Peulic, A.: Thermography based breast cancer detection using texture features and minimum variance quantization. EXCLI J. **13**, 1204–1215 (2014)

11. Golestani, N., Etehad Tavakol, M., Ng, E.Y.K.: Level set method for segmentation of infrared breast thermograms. EXCLI J. **13**, 241–251 (2014)

12. Pramanik, S., Bhowmik, M.K., Bhattacharjee, D., Nasipuri, M.: Segmentation and analysis of breast thermograms for abnormality prediction using hybrid intelligent techniques. In: Hybrid Soft Computing for Image Segmentation, pp. 255–289. Springer (2016)

13. Chan, T., Vese, L.: Active contours without edges. IEEE Trans. Image Proc. **10**, 266–277 (2001)

14. Osher, S., Fedkiw, R.: Level set methods and dynamic implicit surfaces. Appl. Math. Sci. **153**, 1–82, 119–124 (2002)

15. Cheng, L., Yang, J., Fan, X., Zhu, Y.: A generalized level set formulation of the Mumford-Shah functional for brain MR image segmentation. In: IPMI 2005. LNCS 3565, pp. 418-430 (2005)

16. Gupta, A.: Groundwork of Mathematical Probability and Statistics, 3rd edn. Academic Publishers, Calcutta, India (1995)

17. Pramanik, S., Bhattacharjee, D., Nasipuri, M.: Texture analysis of breast thermogram for differentiation of malignant and benign breast. In: IEEE International Conference on Advances in Computing, Communications, and Informatics (2016)

18. Silva, L.F., Saade, D.C.M., Sequeiros-Olivera, G.O., Silva, A.C., Paiva, A. C., Bravo, R.S., Conci, A.: A new database for breast research with infrared image. J. Med. Imaging Health Inform. **4**(1), 92–100(9) (2014)

19. Cardenes, Ruben, de Luis-Garcia, Rodrigo, Bach-Cuadra, Meritxell: A multidimensional segmentation evaluation for medical image data. J. Comput. Methods Programs Biomed. **96**(2), 108–124 (2009)

20. Acharya, U.R., Ng, E.Y.K., Tan, J.H., Sree, S.V.: Thermography based breast cancer detection using texture features and support vector machine. J. Med. Syst. **36**(3), 1503–1510 (2012)

VQ/GMM-Based Speaker Identification with Emphasis on Language Dependency

Bidhan Barai, Debayan Das, Nibaran Das, Subhadip Basu and Mita Nasipuri

Abstract The biometric recognition of human through the speech signal is known as automatic speaker recognition (ASR) or voice biometric recognition. Plenty of acoustic features have been used in ASR so far, but among them Mel-frequency cepstral coefficients (MFCCs) and Gammatone frequency cepstral coefficients (GFCCs) are popularly used. To make ASR language and channel independent (if training and testing microphones and languages are not same), i-Vector feature and unwanted variability compensation techniques like linear discriminative analysis (LDA) or probabilistic LDA (PLDA), within-class covariance normalization (WCCN) are extensively used. At the very present days, the techniques for modeling/classification that are used are Gaussian mixture models (GMMs), vector quantization (VQ), hidden Markov model (HMM), deep neural network (DNN), and artificial neural network (ANN). Sometimes, model-domain normalization techniques are used to compensate unwanted variability due to language and channel mismatch in training and testing data. In the present paper, we have used maximum log-likelihood (MLL) to evaluate the performance of ASR on the databases(DBs), namely ELSDSR, Hyke-2011, and IITG-MV SR Phase-I & II, based on MFCCs and VQ/GMM where the scoring technique MLL is used for the recognition of speakers. The experiment is carried out to examine the language dependency and environmental mismatch between training and testing data.

Keywords MFCC · GFCC · VQ/GMM · i-Vector · PLDA · SVM · MLL score

B. Barai (✉) · D. Das · N. Das · S. Basu · M. Nasipuri
Jadavpur University, Kolkata 700032, India
e-mail: bidhanb@research.jdvu.ac.in

D. Das
e-mail: debayan.157@gmail.com

N. Das
e-mail: nibaran@cse.jdvu.ac.in

S. Basu
e-mail: subhadip@cse.jdvu.ac.in

M. Nasipuri
e-mail: mnasipuri@cse.jdvu.ac.in

© Springer Nature Singapore Pte Ltd. 2019
R. Chaki et al. (eds.), *Advanced Computing and Systems for Security*,
Advances in Intelligent Systems and Computing 883,
https://doi.org/10.1007/978-981-13-3702-4_8

1 Introduction

Speaker Recognition (*SR*) is a branch of science and technology for analyzing characteristics of speech of speakers, which leads to uniquely identify individuals by means of speech waveform. SR was first introduced by Pruzansky et al. [20]. SR has a broad range of applications like access control, security systems, criminal tracking, forensic and many more. There are two types of SR: (a) *closed-set* and (b) *open-set*. In this paper, the result and analysis of closed-set SR is presented. If it is known that the unknown speaker will always be one of the enrolled speakers, then it is called closed-set SR and if the unknown speaker may be one out of the enrolled speakers, then it is called open-set SR. The SR is also divided in two types: (a) *text-dependent* and (b) *text-independent*. If the contents of speech is same for *training* and *testing/classification* then it is called text-dependent SR and if the contents are different, it is called text-independent SR. The SR has two basic steps called *feature extraction* and *modeling/classification*. To measure the performance (accuracy) of SR, *testing* step is required. The performance of SR is measured by the percentage of correctly classified speakers from the testing data of all the speakers.

The SR is further divided into two types: (a) *speaker identification* (*SI*) and (b) *speaker verification* (*SV*). In SI, the test speaker is compared with all the enrolled speakers. For each speaker, a statistical distance (or score) from the unknown speaker is calculated and the minimum distance classifies unknown speaker as identified speaker. In SV, the test speaker claims her/his identity by triggering her/his appropriate speaker template/model to compare with the test speaker and a value p is calculated for the comparison with a threshold value θ which is calculated from the knowledge of train and test data. If $p \geq \theta$, the claim is accepted, otherwise rejected. The SR problem is basically a *pattern recognition* (*PR*) problem. The described method(s) is one of the methods for solving a PR problem. A combined block diagram of SI and SV is shown in Fig. 1.

We conduct SI experiment using *Vector Quantization* (*VQ*) to reduce the number of vectors in feature space. Reynolds et al. mentioned in [22] that MFCCs provide the best result with *Gaussian Mixture Model* (*GMM*)-based SR in noiseless (clean speech) environment keeping recording device same for training and testing data. Hence we choose MFCC feature in feature space. However for environmental and channel mismatch, GFCCs and *i-Vector* provide best results for SI [8, 12, 27]. An interesting result found if training and testing data differ by language. In this case, the accuracy of SI degrades substantially. But if there is a device or environmental mismatch, then the performance of SR degrades more than language mismatch even sometimes degrades drastically. We conducted the experiment with IITG-MV SR (Indian Institute of technology multivariate speaker recognition) [10], Hyke-2011 [21] and ELSDSR [7] bases and the results are reported and analyzed. But to see the effect of language mismatch, IITG-MV SR database is considered as other two databases do not contain data with language variation.

To deploy SR technology over practical application areas, we must combat following difficulties:

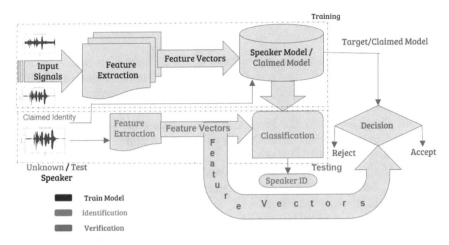

Fig. 1 A combined block diagram of SI and SV

- Availability of speech waveform (utterances) with sufficient duration for training and testing the ASR system.
- The accuracy of SR is highly dependent on language and environment [1]. It is very difficult to collect data for training and testing in same language and environment.
- The device (recording media like microphone) mismatch between training and testing is another factor for the large degradation of accuracy.
- The data to design an ASR system must be reduced as it hard to accumulate and annotate large amount of speech data [2].
- The speech waveform may contain additive noise, convolution noise, and reverberant noise [28].

The above difficulties could be overcome by the filtering the speech signal, feature domain transformation and/or feature, model and score-domain normalization techniques. But the performance (or accuracy) of SR for these conditions is not so up to the mark that it could be applied for real-life or forensic application. However, at present days an array of features, feature extraction techniques and methodologies are developing to combat with the above difficulties. Since the above methods and techniques must go through a number of steps, the components of SR system are increased so as the time of SR evolution process. The example of such a technique is *i-vectors* [13, 25] which have been the state-of-the-art technique over the last few years. Though the present advancement of *Deep Learning (DL)* provides improvement in SR using i-vectors technique, use of those are computationally expensive. Javier et al. [9, 25] proposed vector representation of speech signal using both *GMM and Restricted Boltzmann Machines (RBM)* to reduce the computational load for vector extraction process and transformed the speech signal into a low-dimensional *GMM-RBM vectors* rather than high-dimensional i-vectors.

2   Feature Extraction

The primary and fundamental step to solve an SR problem is *extraction of suitable features*. It transforms raw and physical waveform into the feature space. The most popular features that are used in SI are *Gammatone Frequency Cepstral Coefficient* (GFCC) [27], *Linear Predictive Cepstral Coefficient* (LPCC), *Mel-frequency Cepstral Coefficient* (MFCC), *Perceptual Linear Predictive Cepstral Coefficient* (PLPCC) [23], combination of *MFCC and phase information* [17], *Modified Group Delay Feature* (MODGDF) [16], *i-Vector* [12], *Mel Filter Bank Energy-Based Slope Feature* [15], and *Bottleneck Feature of DNN* (BF-DNN) [26]. In our experiment, only MFCC feature is used to analyze whether SI is language-dependent and experimental results of clean speech and noisy speech for language matching condition are reported. Every enrolled speaker has a set of MFCC vector in feature space upon which GMM is built. Now we shall describe the methods of extracting MFCC features from the speech waveform. Suppose the training data (speech waveform) of S speakers are given for enrollment. These data are transformed into a measurable space, using digital signal processing, where each speaker can be identified uniquely. This measurable space is known as *feature space* and process of transformation is called *feature extraction*. Mathematically, feature extraction is a map $\mathbf{X} = \mathbf{f}(\mathbf{Y})$, by which sample(s) $\mathbf{Y}$ in a space $\Omega_{\mathbf{Y}}$ is transformed into p-dimensional feature space $\Omega_{\mathbf{X}}$ [18] and produces the feature vectors $\mathbf{X}$'s in feature space. Till the present day, a plenty of features are used in ASR but among those, MFCCs and GFCCs are normally preferred by the research community. After feature extraction step, each speaker is represented by a set of feature vectors, $\mathcal{X} = \{\mathbf{x}_{\tau} \in \mathbb{R}^{D} : \tau \in [1, N]\}$, where N be the number of feature vectors in feature space and D be the dimension of feature space.

2.1   MFCC Computation

The complete block diagram to extract MFCC vectors from the speech waveform is shown in Fig. 2.

- **Pre-emphasis**: In this process of time domain, the amplitude of high frequencies of the frequency domain of speech waveform is amplified because higher frequencies provide more speaker specific information than lower frequencies. A *High-Pass Filter* (*HPF*), through which the entire speech waveform is passed, is used to perform this operation. This operation on the signal $s(n)$ (speech waveform) in time domain is given by the simple equation, $\tilde{s}(n) = s(n) - \alpha s(n-1)$ *for* $\alpha \in (0.9, 1)$ Generally, the value of pre-emphasis factor, α is set to 0.97. Here $\tilde{s}(n)$ is the output of the HPF.
- **Short-Time Fourier Transform (STFT)**: For STFT of $x(n)$, where $x(n)$ be a frame for the short-time period, a window function $win(n)$ is chosen. The window function $win(n)$ is defined by

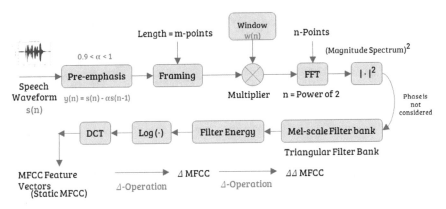

Fig. 2 A diagram to represent different computational steps of MFCC feature vector

$$win(n) = \beta - \gamma \cos(\frac{2\pi n}{\mathcal{L} - 1}) \tag{1}$$

where $\mathcal{L}$ is the window length and $\gamma = 1 - \beta$. For Hanning window, $\beta = 0.5$ and $\gamma = 0.5$ and for Hamming window, $\beta = 0.54$ and $\gamma = 0.46$. The other window functions like Gaussian, Welch, Blackman, Kaiser, and Bartlett could also be applied. At first, we perform windowing operation and then *Discrete Fourier Transform (DFT)* of windowed speech waveform is computed. The equation of this operation given by $X(k) = \mathcal{FFT}\{x(n)win(n)\} \ for \ 0 \leq k \leq N$. To enable *FFT*, we are required to make frame length as power of 2. Hence zeros are padded with the frame to make frame length a nearest power of 2 if $\mathcal{L}$ is not the power of 2, otherwise zero padding is not required. We take $\mathcal{L} = 400$ in the experiment. Hence, for FFT computation, we pad 112 zeros to make frame length is $N = 512 = 2^9$ which is clearly the power of 2.

- **Magnitude Spectrum**: With the help of equation $S(k) = |X(k)|^2 \ for \ 0 \leq k \leq N$, we compute the squared magnitude spectrum which is called as *Power Spectrum*.
- **Mel-Scale Filter Bank (MFB)**: Now we shall compute MFB. To do so we set B number of overlapping triangular filters in the interval $[m_{min}(f), m_{max}(f)]$ in mel scale. Every filter has uniform spacing in *mel* scale and nonuniform spacing in linear (Hz) scale. The overlaps occur at the center frequency (in *mel*). The relation between Mel and Linear (Hz) scale is given by

$$m(f) = 1127 log_e(1 + \frac{f}{700}) \tag{2}$$

where f in Hz and $m(f)$ in *mel*. The filter bank is computed in the frequency range $[f_{min}, f_{max}]$. It is known that human vocal chord cannot produce frequencies more that 5500 Hz. Hence we choose $f_{max} = 5500$ Hz and $f_{min} = \{0, 300\}$ Hz. Here, values of f_{min} and f_{max} are represented by $m_{min}(f)$ and $m_{max}(f)$, respectively, in *mel* scale. Each filter in filter bank is recognized by three frequencies: start, center and end, i.e., $m_s(f)$, $m_c(f)$ and $m_e(f)$, respectively. We can compute f_s, f_c, and f_e using inverse operation of equation (2) given by $f = 700(e^{\frac{m(f)}{1127}} - 1)$ Hz where f in Hz and $m(f)$ in mel. Next, we map the frequencies f_s, f_c, and f_e with the nearest FFT index numbers given by f_{bin}^s, f_{bin}^c, and f_{bin}^e, respectively, known as FFT bins, using

$$f_{bin}^\theta = floor(\frac{(N+1).f_\theta}{bw}), \qquad \theta \in \{s, c, e\} \tag{3}$$

In above the equation, $bw = f_{max} - f_{min}$ is the bandwidth of each filter. At center bin, the filter's amplitude is $f_{bin}^c = 1$ which is maximum and the amplitudes at start and end bins are, $f_{bin}^s = 0$ and $f_{bin}^e = 0$ which are minimum. The amplitudes of each filter are computed as follows:

$$H_m(k) = \begin{cases} 0 & \text{if } k < f_{bin}^s \\ \frac{k - f_{bin}^s}{f_{bin}^c - f_{bin}^s} & \text{if } f_{bin}^s \leq k \leq f_{bin}^c \\ \frac{f_{bin}^e - k}{f_{bin}^e - f_{bin}^c} & \text{if } f_{bin}^c \leq k \leq f_{bin}^e \\ 0 & \text{if } k > f_{bin}^e \end{cases} \tag{4}$$

- **Filter Energy**: The filter bank consisting B filters is superimposed over the power spectrum $S(k)$. Within each filter, amplitudes of filter are multiplied by the corresponding points of $S(k)$ and summed up all the products to get the *filter energy*, denoted by $\{\tilde{S}(k)\}_{k=1}^{k=B}$. We compute *log energies* as $\{log(\tilde{S}(k))\}_{k=1}^{k=B}$. Now we are required to perform *Discrete Cosine Transform* (DCT) of this *log energy* frame to compute MFCC vector.
- **DCT**: The MFCC vector is computed by performing DCT on the log-energy frame by

$$C_n = \sum_{\kappa=1}^{D}(\log \tilde{S}(\kappa))cos(n(\kappa - \frac{1}{2})\frac{\pi}{D}), \tag{5}$$

where $n \in [1, D]$. The MFCC vector C_n has the dimension $D = B$.
- **Liftering**: If C_n is a cepstral coefficient, and w_n is a lifter, then $C_n^l = w_n C_n$ is a liftered cepstral coefficient, where w_n is lifter, defined as:

(i) **Linear Lifter**: $w_n = n$
(ii) **Sinusoidal Lifter**: $w_n = 1 + \frac{D}{2}sin(\frac{n\pi}{2})$
(iii) **Exponential Lifter**: $w_n = n^s e^{-\frac{n^2}{2\tau^2}}$

where s and τ are constants. Their typical values are $\tau = 5$ ans $s = 1.5$. In case of device mismatch, sinusoidal liftering of MFCC's shows significant improvement in accuracy of SR.
- **ΔMFCC and $\Delta\Delta$MFCC Feature**: Δ is known as *velocity* and $\Delta\Delta$ MFCC features is *acceleration*, which are also known as dynamic features. The Δ MFCC feature bears the information of power spectral envelop only. However, speech waveform

would also has information about the dynamics in MFCC feature space, i.e., trajectories of the static MFCC coefficients over time. If the variability among inter-speakers is small for MFCC feature, then velocity and acceleration coefficients increases the inter-speaker variability and decreases the intra-speaker variability. Generally, if Δ and $\Delta\Delta$ features are considered, then $D = 38$ dimensional feature vector is formed by taking $12 - MFCC$, $13 - \Delta MFCC$, and $13 - \Delta\Delta MFCC$ and concatenating. To calculate $\Delta MFCC$, the following formula is used

$$dc_n = \frac{\sum_{r=1}^{J} r(c_{n+r} - c_{n-r})}{2\sum_{r=1}^{J} r^2} \tag{6}$$

Here we must take static MFCC coefficients slightly greater than $D = 13$ (depending on the value of J) because few extra points are needed to calculate Δ feature. A typical value for J is 2 ($J = 1$ is also possible). $\Delta\Delta$ can be calculated by applying Eq. (6) on Δ feature dc_n. In our experiment, we take only static MFCC coefficients with $D = 13$ because in our databases, namely $IITG - MVSR$, $Hyke - 2011$, and $ELSDSR$, inter-speaker variability is significantly large.

- **Cepstral Mean Subtraction (CMS)**: CMS translates the mean MFCC vector to the origin in the feature space to remove the channel variability. For CMS, we compute the mean (μ) MFCC vector of the set of MFCC vectors. Now, this mean vector is subtracted from each of the MFCC vectors in feature space to form new MFCC vectors. If $\mathbf{x}_i$ is the new feature vector after subtracting μ from the original feature vector $\mathbf{x}'_i$, then $\mathbf{x}_i = \mathbf{x}'_i - \mu$. If CMS is not carried out then we denote original feature vector by $\mathbf{x}_i$ in the feature space upon which GMM is built otherwise GMM is built upon mean-subtracted feature space.

Sometimes in order to improve the robustness of speech waveform, the above signal processing technique (used for feature extraction) is combined with another signal processing system to mimic some aspect of the human auditory system for feature extraction. An example of such a feature is *Inner Hair Cell Coefficients (IHC)* feature. Meddis Auditory Periphery model is used to extract this feature. Indeed, a Gammatone filter bank is combined with an IHC model [19]. Though the fundamental frequency-related features like *pitch* and *formants* are very old state-of-the-art feature [6] for SR, they are still in use in combination with GFCC, MFCC, and IHC features to improve SR performance.

3 Modeling/Classification

In Sect. 2, every speaker's speech data are transformed into the feature space. The set of feature vectors of every speaker $\mathcal{X}$ forms the *template (voice-print)* of that speaker. Every speaker is represented by a statistical model in the model-based classifier. A model is represented by a set of model parameters $\boldsymbol{\lambda}$ which is calculated using feature vectors for training data.

3.1   Vector Quantization (VQ)

VG is the process where the feature space is subdivided into multiple numbers of non-overlapping regions and for each region, a *reconstruction vector* is calculated. Every reconstruction vector is used to code the data points (or vectors) in that region. The set of all these reconstruction vectors called a *code book* [18]. A system by which a codebook is computed is known as vector quantizer. A Voronoi or nearest neighbor (NN) quantizer provides minimum encoding distortion. Generally, *LBG algorithm* is used to compute a Voronoi quantizer to condense the set of N vectors $\mathcal{X}$ and represents a speaker by a set C_i, of K reconstruction vectors (also called code vectors) known as code book [18, 24] where $K \ll N$ and $i = 1, 2, 3, \dots, S$. This means that every speaker is modeled by a codebook of K vectors and there will be S code books in the model domain for S speaker. A typical value of K is 1024 for a VQ classifier.

3.1.1   Classification

Let *Euclidean distance* (distortion) between the *D-dimensional* vectors $\mathbf{x}$ and $\mathbf{y}$ is denoted by $d(\mathbf{x}, \mathbf{y})$. The codebook of ith speaker is C_i where $\{i = 1, 2, 3, \dots, S\}$ is known from ith speaker's training data. Suppose that the unknown speaker's data (test data) are transformed into feature space giving N feature vectors. Then the minimum distances with respect to each code book are accumulated (summed up) across all the N test feature vectors. The average distance (distortion) of test speaker with respect to ith codebook (speaker) is

$$dC_i = \frac{1}{K} \sum_{l=1}^{K} \min_{1 \le j \le N} d(\mathbf{x}_l, \mathbf{y}_j^i) \tag{7}$$

Here $\mathbf{y}_j^i$ is the jth code vector of enrolled ith speaker and $\mathbf{x}_l$ is the lth vector of test speaker. Indeed, dC_i is the distance measure between codebook of ith speaker and feature set of unknown (or test) speaker. Now dC_i is measured for all the S speakers and minimum distance provides the unknown (test) speaker as the classified speaker, i.e., the recognized speaker is given by $\hat{S} = \arg\min_{i \in S}(dC_i)$ where S is the set of speakers and $\hat{S}$ is the recognized speaker.

3.2   Gaussian Mixture Model (GMM) and VQ/GMM

Let for ith speaker, the speech data is transformed into the $MFCC$ feature space of K feature vectors of dimension D, viz $\mathcal{X} = \{\mathbf{x}_\tau \in \mathbb{R}^D : 1 \le \tau \le K\}$. The GMM for ith speaker, λ_i, is defined by the sum of M component D-variate weighted Gaussian

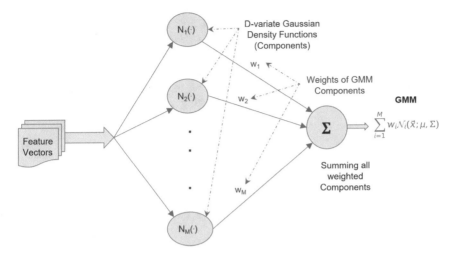

Fig. 3 A block diagram of GMM

densities [22], where mixture *weights* w_j $\{j = 1\ to\ M\}$ must satisfy $\sum_{j=1}^{M} w_j = 1$. Hence, M component GMM model λ_i is given by

$$p(\mathbf{x}_t|\lambda_i) = \sum_{j=1}^{M} w_j \mathcal{N}(\mathbf{x}_t; \boldsymbol{\mu}_j, \boldsymbol{\Sigma}_j) \quad and \quad \sum_{j=1}^{M} w_j = 1 \tag{8}$$

where $\mathcal{N}\ (\mathbf{x}_t; \boldsymbol{\mu}_j, \boldsymbol{\Sigma}_j)$ with $\{j = 1\ to\ M\}$ are D-variate Gaussian density functions given by

$$\mathcal{N}(\mathbf{x}_t; \boldsymbol{\mu}_j, \boldsymbol{\Sigma}_j) = \frac{1}{(2\pi)^{D/2}|\boldsymbol{\Sigma}_j|^{1/2}} e^{-\frac{1}{2}(\mathbf{x}_t - \boldsymbol{\mu}_j)^T \boldsymbol{\Sigma}_j^{-1}(\mathbf{x}_t - \boldsymbol{\mu}_j)} \tag{9}$$

with mean vector $\boldsymbol{\mu}_j \in \mathbb{R}^D$ and covariance matrix $\boldsymbol{\Sigma}_j \in \mathbb{R}^{D \times D}$. $(\mathbf{x}_t - \boldsymbol{\mu}_j)^T$ represents the transpose of vector $(\mathbf{x}_t - \boldsymbol{\mu}_j)$. The GMM model for ith speaker λ_i is parameterized by weight w_j, mean vector $\boldsymbol{\mu}_j$, and covariance matrix $\boldsymbol{\Sigma}_j$, i.e., $\lambda_i = \{(w_j, \boldsymbol{\mu}_j, \boldsymbol{\Sigma}_j) : 1 \leq j \leq M\}$

A block diagram for GMM is shown in Fig. 3. In this figure, $\mathcal{N}(\mathbf{x}_t; \boldsymbol{\mu}_j, \boldsymbol{\Sigma}_j)$ is represented as $\mathcal{N}_j(\cdot)$ for $j = 1, 2, 3, \dots, M$.

Maximum-Likelihood (ML) Parameter Estimation (MLE): After assuming every speaker has the form of GMM, we are required to calculate model parameters λ_i for all the speaker, i.e., for $i = 1, 2, 3, \dots, S$. To do so, we make use of EM algorithm. With the help of EM algorithm, we iteratively reestimate the parameters $w_j, \boldsymbol{\mu}_j, \boldsymbol{\Sigma}_j$) after initialization. The *maximum likelihood* (*ML*) [4] value is computed by

$$p(\mathcal{X}|\lambda_i) = \prod_{\tau=1}^{K} p(\mathbf{x}_\tau|\lambda_i) \tag{10}$$

The EM algorithm starts with an initial model λ_{init} and reestimate a new model λ in such a way that it gives $p(\mathcal{X}|\lambda) \geq p(\mathcal{X}|\lambda_{init})$. Now we are required to estimate model parameters. To do so, using k-means clustering algorithm, we initialize the mean vector $\boldsymbol{\mu}_j$. Using this mean vector, we initialize covariance matrix $\boldsymbol{\Sigma}_j$. w_j is set to $1/M$ as its initial value for all j's. Now all the parameters are reestimated at each EM iteration in accordance with the following equations to get the new speaker model λ_{new}.

$$w_j = \frac{1}{K} \sum_{\tau=1}^{K} \mathcal{P}(j|\mathbf{x}_\tau, \lambda_i) \tag{11}$$

$$\boldsymbol{\mu}_j = \frac{\sum_{\tau=1}^{K} \mathcal{P}(j|\mathbf{x}_\tau, \lambda_i)\mathbf{x}_{tau}}{\sum_{\tau=1}^{K} \mathcal{P}(j|\mathbf{x}_\tau, \lambda_i)} \tag{12}$$

$$\boldsymbol{\Sigma}_j = \frac{\sum_{\tau=1}^{K} \mathcal{P}(j|\mathbf{x}_\tau, \lambda_i)(\mathbf{x}_j - \boldsymbol{\mu}_j)(\mathbf{x}_j - \boldsymbol{\mu}_j)^T}{\sum_{\tau=1}^{K} \mathcal{P}(j|\mathbf{x}_\tau, \lambda_i)} \tag{13}$$

for each i. The iteration continues if a suitable stopping criteria does not hold. In covariance matrix $\boldsymbol{\Sigma}_j$, all off-diagonal elements are set to zero and we consider only diagonal elements. The probability $\mathcal{P}(j|\mathbf{x}_\tau, \lambda_i)$ for each i is given by

$$\mathcal{P}(j|\mathbf{x}_\tau, \lambda_i) = \frac{w_j \mathcal{N}(\mathbf{x}_\tau; \boldsymbol{\mu}_j, \boldsymbol{\Sigma}_j)}{\sum_{j=1}^{M} w_j \mathcal{N}(\mathbf{x}_\tau; \boldsymbol{\mu}_j, \boldsymbol{\Sigma}_j)} \tag{14}$$

VQ/GMM: In *VQ/GMM*, at first, the training (or enrolled) data are reduced by VQ and then the GMM is built upon the reduced data. Generally, the VQ/GMM method is applicable for very large number of feature vectors in the feature space. Here it is assumed that the reduced vectors in codebook (code vectors) follow the GMM. However, if the number of feature vectors in feature space is less (say, 3000 feature vectors), then only GMM is used rather using VQ/GMM. In our experiment, we use codebook of 512 code vectors, over which we build the GMM.

Classification: *SI using MLL Scoring technique.*
We assume there are S speakers in the database represented by $\mathcal{S} = \{1, 2, \ldots, S\}$ and their GMMs are $\lambda_1, \lambda_2, \lambda_3, \ldots, \lambda_S$, respectively. Now, we are required to compute the speaker's GMM model with the maximum *posteriori* probability for the set of feature vectors $\mathcal{X}$ of each test speaker. Using minimum error Bayes' decision rule, the identified speaker is evaluated by

$$\hat{S} = \arg\max_{\kappa \in \mathcal{S}} \left(\frac{p(\mathcal{X}|\lambda_\kappa)Pr(\lambda_\kappa)}{p(\mathcal{X})} \right) \tag{15}$$

Here we make the assumption that speakers are equally likely and observing $p(\mathcal{X})$ is same for every speaker, we can find that $Pr(\lambda_\kappa)$ and $p(\mathcal{X})$ are same for all speakers.

Hence Eq. (15) becomes $\hat{S} = \arg\max_{\kappa \in \mathcal{S}} (p(\mathcal{X}|\lambda_\kappa))$. Now we assume that the vectors are independent and taking logarithm of Eq. (10), the computation becomes as follows:

$$\hat{S} = \arg\max_{\kappa \in \mathcal{S}} (\sum_{\tau=1}^{K} log(p(\mathbf{x}_\tau | \lambda_\kappa))) \tag{16}$$

Here $\hat{S}$ is the identified speaker and κth speaker's LL score is given by $\sum_{\tau=1}^{K} log(p(\mathbf{x}_\tau|\lambda_\kappa))$. The identified speaker $\hat{S}$ has the maximum log-likelihood (MLL) score.

4 Experimental Setup and Results with Discussion

Names of the some standard popular databases used for SR evaluation are summarized in Table 1.

In the present experiment, the three databases, namely IITG multi-variability SR (IITG-MV SR Phase I & II) database, ELSDSR, and Hyke-2011 are used for the SR evaluation in both environmental match and mismatch conditions. English and/or Indian regional languages like Hindi and Tamil are used in these speech databases. The IITG-MV SR Phase-I has the utterances that are recorded by the different devices, namely Digital Voice Recorder (DVR-D01), Headset (HCL-H01), Tablet PC (T01), Nokia 5130c mobile (online recording—M01), and Sony Ericsson W350i mobile (offline recording—M02) in noisy office environment. Whereas the IITG-MV SR Phase-II has the utterances that are recorded by the devices such as Digital Voice Recorder (DVR—D01), Headset (H01), Tablet PC (T01), Nokia 5130c mobile (online recording—M03), and Sony Ericsson W350i mobile (offline recording—M04) in noisy multi-environment condition (in Laboratory and Hostel room). Here online means recorded through telephone network and offline means mobiles are used as recording device. Every recording device has speech waveform for training and testing in the both phases. The experiment is carried out in "no device mismatch" condition except for M01 is replaced by M03 and M02 is replaced by M04 in Table 2 in testing to examine device mismatch. Also, this noisy environment is same for both training the GMM and testing the ASR system. However, ELSDSR and Hyke-2011 contain clean utterances, i.e., noise level is very low and are recorded with same microphone to examine the accuracy in matched device condition. The sampling frequency for D01, H01, and T01 are 16 kHz whereas for M01, M02, M03, M04, ELSDSR, and Hyke-2011 are 8 kHz. In experiment frames of size about 25 ms with 17 ms overlap, i.e., frame shift of $(25-17) = 8$ ms for 16 kHz speech signal are chosen, whereas for 8 kHz speech signal 50 ms frame and about 34 ms overlap, i.e., frame shift of $(50 - 34) = 16$ ms are chosen. The pre-emphasis factor $\alpha = 0.97$ is considered, and a 512-point FFT algorithm is used. For MFB, maximum and minimum linear frequencies are $f_{min} = \{0, 300\}$ Hz and $f_{max} = 5500$ Hz. The number of

Table 1 The list of some important databases of speech/speaker recognition along with their characteristics

DB name	Language	Rec. equipment(s)	Category of utterance				# speaker	Made by	Vendor
			Sentence	Word	Digit	Spontaneous			
Russian speech database	RUS	Mic	Yes	No	No	No	89	STC	ELRA
IITG-MV SR (Phase I, II, III & IV)	Eng. (IND) + 13 Regional Languages	Mobile(2), DVR, Tablet PC, Headset mic	Yes	No	No	Yes	400	Indian Institute of Technology, Guwahati	IITG
TIMIT/NTIMIT	Eng.	mic, tel	Yes	No	No	No	630	MIT, TI, SRI	LDC
XM2VTS	Eng.	Mic, video	No	No	Yes	No	295	Univ. of Surrey	Univ. of Surrey
Hyke-2011	Eng. (IND)	Mic	Yes	Yes	Yes	NO	100	Microsoft	Microsoft
ELSDSR	Eng.	Solid State Rec.	Yes	Yes	No	No	22	IMM,DAN	IMM
NIST SRE	Eng.	Tel, mic	Yes	Yes	Yes	Yes	2000+	NIST	NIST
Polycost	Eng.	Tel	Yes	No	Yes	Yes	134	COST250	ELRA
EUROM-1	Danish	mic	Yes	No	Yes	No	60	Tele Danmark, CPK	ELRA
SWB	Eng.	Mic	No	No	Yes	No	138	TI	TI
SITW	Eng.	Multi-media	Yes	Yes	No	No	299	MIT-LL	LDC
VoxCeleb	Eng.	Multi-media	Yes	Yes	Yes	Yes	1251	–	–
MIT Mobile	Eng.	mic, tel	Yes	Yes	Yes	Yes	88	MIT	MIT

Table 2 Performance evaluation on databases IITG-MV SR, Hyke-2011, and ELSDSR for 5 EM iteration and 512 VQ clusters under same environment, language, and recording device

DB name & Rec. equipment		Utterance type	# speakers	# Gaussian compts.	Starting frequency (f_{min})	Dim. of MFCC (D)	Testing time (s)	Accuracy (%)
IITG-MV SR	DVR (D01)	Noisy	100	16	0	13	3638	96
				16	300	13	3287	95
				32	0	13	5119	96
				32	300	13	5336	**96**
	Headset (H01)	Noisy	100	16	0	13	4010	70
				16	300	13	4239	89
				32	0	13	5871	76
				32	300	13	5801	**93**
	Tablet PC (T01)	Noisy	100	16	0	13	3922	90
				16	300	13	3513	89
				32	0	13	5433	**91**
				32	300	13	5129	89
	Nokia 5130c (M01)	Noisy	100	16	0	13	2570	94
				16	300	13	2741	**95**
				32	0	13	4434	92
				32	300	13	4806	94
	Sony Ericsson W350i (M02)	Noisy	100	16	0	13	2824	86
				16	300	13	2641	87
				32	0	13	4456	86
				32	300	13	4645	**90**
Hyke-2011	Mic	Clean	83	16	0	13	3660	**100**
				16	300	13	3830	100
				32	0	13	6780	100
				32	300	13	6812	100
ELSDSR	Mic	Clean	22	16	0	13	656	**100**
				16	300	13	714	100
				32	0	13	1292	100
				32	300	13	1367	100

triangular filters in MFB is $B = 26$ which give 26 MFC coefficients among which only first 13 MFC coefficients are chosen to form MFCC feature vector. The accuracy rate for the mentioned databases are shown in Table 2. For VQ, we use 512 clusters, for vector reduction (note that not dimension reduction), upon which GMM is built with the help of 5 EM iteration. In Table 2, we present the recognition performance under matched (language, environment and device) condition.

In the presented SR experiment, a fascinating observation is found in environment mismatch case. In Table 4, the experimental results for environment mismatch case for all the five devices are presented. It is easily observed that there is considerable degradation in performance for the environment mismatch. But performance decreases drastically for M01–M03 and M02–M04 devices due to device mismatch and online–offline condition. The reason for this degradation in performance is due

Table 3 Recognition performance of IITG-MV SR database of all devices for mismatch (language) condition under similar environment and mismatched device case

Rec. devices	# speakers	Training environment	Testing environment	Parameters and constants for MFCC and GMM	Performance (%)
D01	100	Office	Office	$f_{min} = 0$ Hz, $f_{max} = 5500$ Hz, $N_G = 32$, $D = 13$, $B = 26$, $\alpha = 0.97$, $EM\ iteration = 5$, $VQ\ Clusters = 512$, 0th $MFCC$ coefficient C_0 is excluded	63
H01	100				80
T01	100				52
M01	100				56
M02	100				49

to the mismatch in noise level in training and testing data. This different noise level shifts test MFCC vectors from trained MFCC vectors considerably in the scatter plot of train and test data. Here we choose $f_{min} = 0$ Hz, vector dimension $D = 13$, no. of Gaussian components $N_G = 32$, 512 VQ clusters, and 5 EM iteration to create GMM models.

We can observe clearly that the performance of SR degrades heavily for noisy speech when compared with clean speech. This low performance is due to the fact that the frequency spectrum of the signal is distorted considerably by the noise level in the speech signal and this noise level shifts the data in the feature space from the original MFCC vectors. The highest accuracy is observed for $D = 13$ and $N_G = 32$ for all the databases and accuracies degrade beyond this limits. Another fascinating observation is the effect of bandwidth of the filters in the filter bank. We can observe that bandwidth ($f_{max} - f_{min}$) influences the accuracy rate. Language dependency and device mismatch are other important issues for SR are. We observe in Table 3 that the performance degrades when a mismatch of language between train and test data is found. It can be noted that the performance during device mismatch is more drastic on environment mismatch. In Table 3, recognition accuracy is given for the language mismatch in same environment, i.e., noise levels are same in training and testing data. For language mismatch and for devices D01, H01, T01, M01, and M02, the accuracies degrade 33%, 4%, 39%, 36%, 37%, respectively (see Tables 2 and 3). Since for device H01, the accuracy degrades less; then, we can infer that different devices distort speech signal differently as both the language and content of the speech signals are different. However for the devices D01, T01, M01, and M02, the accuracy degrades more. The comparison of Tables 2 and 3, and Tables 2 and 4 shows that the accuracy degradation is more drastic for the environmental and device mismatch than language mismatch.

In the Table 5, the reference works of few popular researchers are given for the quick reference of state-of-the-art methodologies for speaker recognition.

Table 4 Recognition performance for IITG-MV SR database of all devices in environment mismatch condition, i.e., noise levels in train and test data are not same

Rec. devices	# speakers	Training environment	Testing environment	Parameter and constants for MFCC and GMM	Performance (%)
D01	53	Office	Laboratory and Hostel room environment	$f_{min} = 0$ Hz, $f_{max} = 5500$ Hz, $N_G = 32$, $D = 13$, $B = 26$, $\alpha = 0.97$, $EM\ iteration = 5$, $VQ\ Clusters = 512$, 0th $MFCC$ coefficient C_0 is excluded	33.96
H01	53				64.15
T01	53				50
M03	53				18.87
M04	53				22.64

Table 5 A table of reference works of researchers who conducted SR experiments, feature used by them and modeling/classification techniques

Researchers	Feature used	Speaker model	Classification methods
Murthy et al.	MFCC, Δ MFCC, $\Delta\Delta$ MFCC; MFCC with CMN; GD Function (GDF); (MODGDF)	Model-based channel compensation; Synthetic GMM; VQ; VQ-GMM; Variance distribution; Affine transformation of variances; SVM	Euclid distance SVM; Maximum Log-Likelihood (MLE)
Togneri et al.	MFCC with and without temporal derivatives (upto 2nd order); MFCC with CMN and variance normalization	GMM; GMM-SVM; GMM-UBM	MAP; MLL
Schwarz et al.	MFCC; δ MFCC $\Delta\Delta$ MFCC; RASTA filtering; Feature mapping; Segmentation; Feature warping	GMM; GMM/HLDA; Eigen-channel adaptation; GMM-UBM; Adapted GMM	MAP; MLL
Pekhovsky et al.	MFCC with temporal derivatives	GMM; GMM-UBM	Variational Bayesian Model Selection for GMM and GMM-UBM (VB-GMM)

5 Conclusion and Future Scope

SR has a long history of research, over six decades. However, SR research still remains interesting for many researches due to the difficulties arise in SR task. SR using clean speech (noise level is very low) and matched (language, environment, and device) shows very high accuracy rate, but accuracies decrease in mismatch condition. Even environmental and device mismatch influence the accuracy rate very significantly. This is the reason why SR still remains very popular among the researchers. Though, lots of techniques are developed for SR but significant improvement in accuracy rate is still required. We can develop and modify techniques in feature domain, modeling domain, and classification domain. Advancement in machine learning (ML) and deep learning (DL) is providing a significant momentum in SR research. At present days, researchers are heading toward the ML and DL techniques for the SR. The noise removal is a very important issue for SR. The noise can be removed before feature extraction [3, 5, 11] or after feature extraction [14].

Acknowledgements This project is partially supported by the CMATER laboratory of the Computer Science and Engineering Department, Jadavpur University, India, TEQIP-II, PURSE-II and UPE-II projects of Govt. of India.

References

1. Barai, B., Das, D., Das, N., Basu, S., Nasipuri, M.: An ASR system using MFCC and VQ/GMM with emphasis on environmental dependency. In: 2017 IEEE Calcutta Conference (CALCON), pp. 362–366, Dec 2017
2. Barai, B., Das, D., Das, N., Basu, S., Nasipuri, M.: Closed-set text-independent automatic speaker recognition system using VQ/GMM. In: Intelligent Engineering Informatics, pp. 337–346. Springer Singapore, Singapore (2018)
3. Bie, F., Wang, D., Wang, J., Zheng, T.F.: Detection and reconstruction of clipped speech for speaker recognition. Speech Commun. **72**, 218–231 (2015)
4. Dempster, A.P., Laird, N.M., Rubin, D.B.: Maximum likelihood from incomplete data via the EM algorithm. J. R. Stat. Soc. Ser. B (Methodol.) **39**, 1–38 (1977)
5. Dişken, G., Tüfekçi, Z., Saribulut, L., Çevik, U.: A review on feature extraction for speaker recognition under degraded conditions. IETE Tech. Rev. **34**(3), 321–332 (2017)
6. Fant, G.: Acoustic Theory of Speech Production: With Calculations Based on X-Ray Studies of Russian Articulations, p. 2. Walter de Gruyter (1971)
7. Feng, L., Hansen, L.K.: A new database for speaker recognition. Technical report (2005)
8. Garcia-Romero, D., Espy-Wilson, C.Y.: Analysis of i-vector length normalization in speaker recognition systems. Interspeech **2011**, 249–252 (2011)
9. Ghahabi, O., Hernando, J.: Restricted Boltzmann machines for vector representation of speech in speaker recognition. Comput. Speech Lang. **47**, 16–29 (2018)
10. Haris, B.C., Pradhan, G., Misra, A., Prasanna, S., Das, R.K., Sinha, R.: Multivariability speaker recognition database in Indian scenario. Int. J. Speech Technol. **15**(4), 441–453 (2012)
11. Hirszhorn, A., Dov, D., Talmon, R., Cohen, I.: Transient interference suppression in speech signals based on the OM-LSA algorithm. In: International Workshop on Acoustic Signal Enhancement; Proceedings of IWAENC 2012, pp. 1–4. VDE (2012)

12. Kanagasundaram, A., Vogt, R., Dean, D.B., Sridharan, S., Mason, M.W.: I-vector based speaker recognition on short utterances. In: Proceedings of the 12th Annual Conference of the International Speech Communication Association, pp. 2341–2344. International Speech Communication Association (ISCA) (2011)
13. Kanrar, S.: i vector used in speaker identification by dimension compactness. arXiv:1704.03934 (2017)
14. Kheder, W.B., Matrouf, D., Bousquet, P.M., Bonastre, J.F., Ajili, M.: Fast i-vector denoising using map estimation and a noise distributions database for robust speaker recognition. Comput. Speech Lang. **45**, 104–122 (2017)
15. Madikeri, S.R., Murthy, H.A.: Mel filter bank energy-based slope feature and its application to speaker recognition. In: 2011 National Conference on Communications (NCC), pp. 1–4. IEEE (2011)
16. Murthy, H.A., Yegnanarayana, B.: Group delay functions and its applications in speech technology. Sadhana **36**(5), 745–782 (2011)
17. Nakagawa, S., Wang, L., Ohtsuka, S.: Speaker identification and verification by combining MFCC and phase information. IEEE Trans. Audio Speech Lang. Process. **20**(4), 1085–1095 (2012)
18. Pal, S.K., Mitra, P.: Pattern Recognition Algorithms for Data Mining. CRC Press (2004)
19. Paulose, S., Mathew, D., Thomas, A.: Performance evaluation of different modeling methods and classifiers with MFCC and IHC features for speaker recognition. Procedia Comput. Sci. **115**, 55–62 (2017)
20. Pruzansky, S.: Pattern-matching procedure for automatic talker recognition. J. Acoust. Soc. Am. **35**(3), 354–358 (1963)
21. Reda, A., Panjwani, S., Cutrell, E.: Hyke: a low-cost remote attendance tracking system for developing regions. In: Proceedings of the 5th ACM Workshop on Networked Systems for Developing Regions, pp. 15–20. ACM (2011)
22. Reynolds, D.A., Rose, R.C.: Robust text-independent speaker identification using Gaussian mixture speaker models. IEEE Trans. Speech Audio Process. **3**(1), 72–83 (1995)
23. Sapijaszko, G.I., Mikhael, W.B.: An overview of recent window based feature extraction algorithms for speaker recognition. In: 2012 IEEE 55th International Midwest Symposium on Circuits and Systems (MWSCAS), pp. 880–883. IEEE (2012)
24. Soong, F.K., Rosenberg, A.E., Juang, B.H., Rabiner, L.R.: Report: a vector quantization approach to speaker recognition. AT&T Tech. J. **66**(2), 14–26 (1987)
25. Xu, L., Lee, K.A., Li, H., Yang, Z.: Rapid computation of i-vector. In: Odyssey: The Speaker and Language Recognition Workshop, pp. 47–52 (2016)
26. Yamada, T., Wang, L., Kai, A.: Improvement of distant-talking speaker identification using bottleneck features of DNN. In: Interspeech, pp. 3661–3664 (2013)
27. Zhao, X., Wang, D.: Analyzing noise robustness of MFCC and GFCC features in speaker identification. In: 2013 IEEE International Conference on Acoustics, Speech and Signal Processing (ICASSP), pp. 7204–7208. IEEE (2013)
28. Zhao, X., Wang, Y., Wang, D.: Robust speaker identification in noisy and reverberant conditions. IEEE/ACM Trans. Audio Speech Lang. Process. (TASLP) **22**(4), 836–845 (2014)

Natural Language Processing: Speaker, Language, and Gender Identification with LSTM

Mohammad K. Nammous and Khalid Saeed

Abstract Long short-term memory (LSTM) is a state-of-the-art network used for different tasks related to natural language processing (NLP), pattern recognition, and classification. It has been successfully used for speech recognition and speaker identification as well. The amount of training data and the ratio of training to test data are still the key factors for achieving good results, but have their implications on the real usage. The main contribution of this paper is to achieve a high rate of speaker recognition for text-independent continuous speech using small ratio of training to test data, by applying long short-term memory recursive neural network. A comparison with the probabilistic feed-forward neural network has been made for speaker recognition as well as gender and language identification.

Keywords LSTM · Probabilistic neural networks · Speaker identification

1 Introduction and State of the Art

The interaction between computers and human being has always been the subject of interest for researchers in computer science and artificial intelligence. Natural language processing (NLP) is the topic that deals with the intelligent methods of human language processing. The most developing areas within this topic are the

M. K. Nammous (✉)
Fraunhofer Institute for Intelligent Analysis and Information Systems IAIS,
Sankt Augustin, Germany
e-mail: Mohammad.Kheir.Nammous@iais.fraunhofer.de

M. K. Nammous
Faculty of Mathematics and Information Sciences, Warsaw University
of Technology, Warsaw, Poland

K. Saeed
Faculty of Computer Science, Bialystok University
of Technology, Bialystok, Poland
e-mail: k.saeed@pb.edu.pl

© Springer Nature Singapore Pte Ltd. 2019
R. Chaki et al. (eds.), *Advanced Computing and Systems for Security*,
Advances in Intelligent Systems and Computing 883,
https://doi.org/10.1007/978-981-13-3702-4_9

speech recognition, natural language understanding, and natural language generation [1, 2]. Researches in speech recognition differ between text-dependent and text-independent speech recognition; meanwhile, the first could give a high recognition rate for speaker identification [3]; the success rate in the latter was much lower [4]. Therefore, the authors were looking for new methods to keep the recognition performance in a good stand.

In this paper, the domain of speaker identification in text-independent continues speech which is introduced using different techniques for feature extraction and classifications, where the main contribution is to use the LSTM NN to leverage the identification ratio and to make it more stable. The obtained performance of this recursive neural network will be compared with a classical feed-forward neural network in different scenarios. These scenarios are related mainly to the ratio of training to testing data and to other classification problems like gender and language identification.

I-vector and mel-frequency cepstrum coefficients (MFCC) are the most common methods for feature extractions in speech recognition systems [5, 6]. Their applications include different tasks like emotion recognition [7] and age estimation [8]. The mentioned methods have been used for feature extraction; meanwhile for a long time, the Gaussian mixture models (GMMs) have dominated most speech recognition approaches for both feature extraction and classification [9].

Since we are considering the classification part of the recognition process, no background information about the used methods for feature extraction will be presented. The used methods include Burg, TM-eigenvalue algorithm [10], and mel-frequency cepstrum coefficients. Further information is available in earlier publications [3, 4, 10].

During the recent years, the artificial neural networks (ANNs) have become more and more actively used, where different kinds of neural networks have been implemented [11–13]. Recently, the deep neural networks are the newer state-of-the-art method for speaker identification, as well as other feature classifications like age, language, and gender [14–16].

Researches are using different datasets for developing and testing speaker identification and verification algorithms, some obtained from single-speaker environments, others from multi-speaker ones [17]. Few corpuses are available for free like ELS-DSR [18] and SITW [19], and many others still need to be paid for like MIT Mobile [20], SWB [21], and NIST SRE [22]. Many researches also use hybrid corpuses consist of speech signals and visual items for enhancing the recognition rates [23, 24].

2 Database

Authors of this work have worked out their own database. A recorded speech from different online and TV broadcasts in three languages is used: Arabic, English, and Polish; for each language, there are eight speakers. The original recordings contained

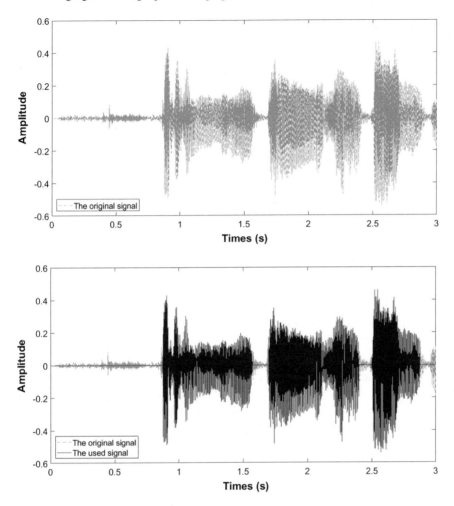

Fig. 1 Original and the obtained signal after eliminating low-energy points from the original signal

mix of speakers' speeches. Separating these speeches into individual files for each speaker is done manually.

The recorded database contains 2 h, 20 min, and 27 s of talks in 44 kHz frequency. The preprocessing step consists of one only step, which aims to eliminate the silence, and thus the input file is scanned to determine the energy levels of the recorded voice and eliminate the low-energy fragments (Fig. 1). After applying this step, the used database will be consisting of 2 h, 7 min, and 22 s.

Table 1 describes the distribution of the database for the three languages and for both genders.

Table 1 Used database

	The complete database (hh:mm:ss)
Arabic	00:38:20
English	00:38:34
Polish	00:50:28
Male	01:25:43
Female	00:41:38
Total	02:07:22

3 The Classification Tools

3.1 Radial Neural Network—RNN

An RBF network has three layers: input, radial, and output layers. The hidden (radial) layer consists of radial units; each actually is modeling a Gaussian response surface. The units will always be sufficient to model any function. The RBF in each radial unit has a maximum of 1 when its input is 0. As the distance between the weight vector and the input decreases, the output increases. Thus, a radial basis neuron acts as a detector that produces 1 whenever the input is identical to its weight vector; additionally, there is a bias neuron, which allows the sensitivity of the radial basis transfer function of the hidden neurons to be adjusted. The standard RBF NN has an output layer radial basis networks may require more neurons than standard feed-forward backpropagation networks, but often they can be designed in a fraction of the time and it takes to train standard feed-forward networks. They work best when many training vectors are available. RBF networks have a number of advantages. First, they can model any nonlinear function using a single hidden layer, which eliminates some design decisions about the number of layers. Second, the simple linear transformation in the output layer can be optimized fully using traditional linear modeling techniques, which are fast and do not pose problems. RBF networks can therefore be trained extremely quickly; training of RBFs takes place in distinct stages. The centers and deviations of the radial units must be set up before the linear output layer is optimized [25].

3.2 Probabilistic Neural Network—PNN

PNN is a feed-forward neural network with a complex structure. It is composed of an input layer, a pattern layer, a summation layer, and an output layer. Despite its complexity, PNN only has a single training parameter. This is a smoothing parameter of the probability density functions (PDFs) which are utilized for the activation of the neurons in the pattern layer. Thereby, the training process of PNN solely requires a single input–output signal pass in order to compute network response. However, only the optimal value of the smoothing parameter gives the possibility of correctness of the model's response in terms of generalization ability [26, 27].

Probabilistic neural network (PNN) is an example of the radial basis function-based model effectively used in data classification problems. It was proposed by Specht [10, 27] and, as the data classifier, draws the attention of researchers from the domain of data mining. It has been widely used for different classification problems [11, 26–28].

3.3 Long Short-Term Memory Recursive Neural Network—LSTM

Long short-term memory network (LSTM) is a typical recurrent neural network, which alleviates the problem of gradient diffusion and explosion. LSTM can capture the long dependencies in a sequence by introducing a memory unit and a gate mechanism which aims to decide how to utilize and update the information kept in memory cell [12, 29, 30].

Currently, the long short-term memory (LSTM) is a state-of-the-art method used for different tasks related to natural language processing (NLP), pattern recognition, and classification tasks [31–34]; it is also widely used for speech and speaker recognition [35, 36].

3.4 The Used LSTM NN Architecture

The used architecture of the LSTM neural network consists of six layers:

- The sequence input layer contains the input feature vector of length depending on the used parameters; in our case, it will contain with 129 values for Burg method, 125 for TM-eigenvalue algorithm, and 23 for MFCC method.
- The LSTM layer with 24 hidden units.

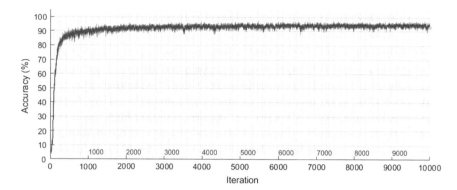

Fig. 2 Training the SLTM NN for speaker identification (MFCC method)

- The dropout layer with dropout possibility 50%.
- Fully connected layer with 24 neurons.
- Softmax layer which applies a softmax function to the input.
- The classification output layer computes the cross entropy loss function, for the k mutually exclusive classes.
- For language and gender identification tasks, a smaller number of neurons have been used in the LSTM and the output layers, because of the reduced number of final categories.

For training this network, the next parameters have been used: epochs: 10,000, batch size: 500, initial learn rate: 0.1, with the stochastic gradient descent with momentum function for updating the weights and biases of the neurons. Sample training process is presented in the next figure (Fig. 2).

4   Experiment Results

Our experiments covered three cases: speaker, language, and gender identification; for each case, there is different number of possible outputs (categories); Table 2 presents the output number for each one.

Table 2 Number of classification's categories

	Speaker Identification	Language Identification	Gender Identification
Number of categories	24 Speakers	3 Languages	2 Genders

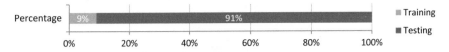

Fig. 3 Overall training to testing ratio, taking 30 s for each speaker as training set

Table 3 Distribution of the used database relating to the languages and genders in case of 30 s training data for each speaker

	The training set (hh:mm:ss)	The testing set (hh:mm:ss)	The complete database (hh:mm:ss)
Arabic	00:04:00	00:34:20	00:38:20
English	00:04:00	00:34:34	00:38:34
Polish	00:04:00	00:46:28	00:50:28
Male	00:08:00	01:17:43	01:25:43
Female	00:04:00	00:37:38	00:41:38
Total	00:12:00	01:55:22	02:07:22

4.1 For 30 s of Speaker Speech as a Training Set

Our baseline experiment will use just 30 s for each speaker to be included in the training set; meanwhile, the rest of the voice signal will be under testing. Moreover, each of both parts will be divided into one-second segments (Fig. 3).

Each part is divided into equal segments with length of one second and the distribution of the used database for the three languages and for genders will be as in Table 3.

4.2 The Motivation, How It Was Before Applying Deep Learning?

In our recent researches [4, 37], two main neural networks have been used for the classification stage: probabilistic and radial neural networks. In this research, we started applying the same methods and comparing the obtained results presented in Table 4.

Table 4 Comparing the recognition ratios for Burg, TM, and MFCC methods, by using two different NNs

Speaker Identification	Recognition rate %	
	Probabilistic NN	Radial NN
Burg's estimation	47%	21%
TM-Eigenvalues	24%	20%
MFCC	72%	26%

Language Identification	Recognition rate %	
	Probabilistic NN	Radial NN
Burg's estimation	96%	92%
TM-Eigenvalues	70%	84%
MFCC	99%	51%

Gender Identification	Recognition rate %	
	Probabilistic NN	Radial NN
Burg's estimation	72%	72%
TM-Eigenvalues	51%	62%
MFCC	88%	57%

Both of probabilistic and radial NN could return accepted good results, with the advancement of the probabilistic one; meanwhile, the radial neural network failed in speaker identification task. Recall that it still has the ability of improving the results obtained by TM-eigenvalue algorithm.

In addition, with the aim of having good classifier for limited training to testing data ratio, these two NNs have limited success. To make a significant improvement, we intend to test the deep learning approach using LSTM. To simplify the case, we will limit next researches into fewer methods, without the need for the details of TM-eigenvalue algorithm for feature extraction and the radial neural networks for classification.

4.3 LSTM as a Main Classifier

Applying long short-term memory could have a significant improvement and can open the door for further investigations and applications; different parameters have

Table 5 Comparing the recognition ratios for Burg and MFCC methods, by using two different NNs

Speaker Identification	Recognition rate %	
	Probabilistic NN	LSTM NN
Burg's estimation	47%	64%
MFCC	72%	89%

Language Identification	Recognition rate %	
	Probabilistic NN	LSTM NN
Burg's estimation	96%	98%
MFCC	99%	100%

Gender Identification	Recognition rate %	
	Probabilistic NN	LSTM NN
Burg's estimation	72%	80%
MFCC	88%	96%

been tested including the number of the layers, number of hidden neurons in the middle layers, and the role of the dropout layer. Simplified general results are given in Table 5.

Using very small amount of training data, LSTM NN could achieve satisfied result for the main task of speaker identification; in other cases, the results were perfect (language identification) and very good for the gender identification task. One important thing is that LSTM NN succeeded in achieving good results for all testified tasks. Even for the cases in which probabilistic NN have not given enough good results, the LSTM NN could clearly increase the recognition rate.

4.4 Increasing the Length of the Segments

Having the same 30 s of each speaker speech as a training data, we will consider dividing the training and the testing data into longer segments. This should extend the information used for all operations and may drive to better results. In the previous experiments, each segment consisted of one-second length, whereas in this experiment we will have 5 s length for each segment. Table 6 presents the obtained results.

Table 6 Comparing the recognition ratios for Burg and MFCC methods, by using two different NNs

Speaker Identification	Recognition rate %	
	Probabilistic NN	LSTM NN
Burg's estimation	46%	65%
MFCC	90%	92%

Language Identification	Recognition rate %	
	Probabilistic NN	LSTM NN
Burg's estimation	95%	99%
MFCC	100%	100%

Gender Identification	Recognition rate %	
	Probabilistic NN	LSTM NN
Burg's estimation	77%	77%
MFCC	93%	99%

The successful rates obtained using LSTM NN have increased a bit; the previous results were already high and increasing the amount of data for each segment increased the success rates more. However, probabilistic NN could not react the same way, and the results were not as good as in the previous case (with smaller segments). The authors think that because of the reduced number of samples, for which PNN was not as good, flexible, and robust as the LSTM NN.

4.5 Would It Be Improved Using More Training Data?

One of the major ratios in recognition tasks is the training to testing data ratio. In points 4.1, 4.2, and 4.3, we presented the results obtained for 30 s of speaker speech as training data, and here we are going to double this amount of training data (the training to testing data ratio will become larger). In general, we are considering two cases in this research: (9/91) and (19/81) training to testing data ratios, which are

Fig. 4 Overall training to testing ratio, taking 1 min for each speaker as training set

Table 7 Distribution of the used database relating to the languages and genders in case of 1 min training data for each speaker

	The training set (hh:mm:ss)	The testing set (hh:mm:ss)	The complete database (hh:mm:ss)
Arabic	00:08:00	00:30:20	00:38:20
English	00:08:00	00:30:34	00:38:34
Polish	00:08:00	00:42:28	00:50:28
Male	00:16:00	01:09:43	01:25:43
Female	00:08:00	00:33:38	00:41:38
Total	00:24:00	01:43:22	02:07:22

already much smaller than the more classical ones (60/40) or (75/25). In the first case (presented in the mentioned points), we used 30 s for training, and in this section, our training dataset will consist of 1 min of speech signal for each speaker, both training and testing set will be divided into 10 s equal samples, which is 10 times longer comparing with the first tested case (experiments 4.1, 4.2) and double the length of the segments used in 4.3. The overall training to testing ratio will be the next (Fig. 4).

And the distribution of the used database for the three languages and for genders will be as shown in Table 7.

The training data will be used to train the networks for the three tested cases (speaker, gender, and language); next the trained networks will be used to testify the recognition rate based upon the training data, and the results are presented in Table 8.

The interpretation of the last case and the comparison with the older ones will be presented in the conclusion section.

Table 8 Comparing the recognition ratios for Burg and MFCC methods, by using two different NNs

Speaker Identification	Recognition rate %	
	Probabilistic NN	LSTM NN
Burg's estimation	58%	72%
MFCC	96%	97%

Language Identification	Recognition rate %	
	Probabilistic NN	LSTM NN
Burg's estimation	99%	100%
MFCC	100%	100%

Gender Identification	Recognition rate %	
	Probabilistic NN	LSTM NN
Burg's estimation	84%	86%
MFCC	94%	98%

5 Conclusions

The authors are not covering the content perspective of understanding languages, rather other NLP tasks and classifications related to the speakers including the spoken language. This type of classifications supports more comprehensive applications in acquiring the right information and applying the related algorithms. The authors are thinking about this project as a pilot test before applying these methods on large-scale datasets, and the novelty of this work is the ability of achieving very good results using small portions of data.

The cause of using the LSTM NN was to improve the recognition rate in the case of text-independent speaker identification. And by using the LSTM, the authors could achieve very good results for all recognition tasks: speaker, language, and gender identification. These networks have given superior ratios in all cases. Their high-end recognition ratio was achieved in combination with the MFCC method and for small classification problems; meanwhile, it could clearly increase the recognition ratio of the other method (Burg's [3], for example) as well.

A special care was given to the usage of very small amount of training data and much smaller amount of information for each sample. The main point was to compare: How each of the neural networks will handle such a difficult task, where using LSTM had a significantly improved the success rate.

Having longer segments (longer used voice samples) increased the information contained in these samples but caused to reduce the number of samples accordingly. In such a case, PNN could not handle such a situation while the LSTM did. The later one was flexible to the situations and conditions we tried. Increasing the training to testing data ration had improved the results as well as using longer samples; however, good results could be achieved even with much smaller amounts of input data and with shorter segments.

The authors believe the presented model is robust enough to be tested now on more complex and much larger datasets, which will be our focus in our next researches.

Acknowledgements This work was supported by grant S/WI/3/2018 from Bialystok University of Technology and funded with resources for research by the Ministry of Science and Higher Education in Poland.

References

1. Jurafsky, D., Martin, J.H.: Speech and Language Processing, 2nd edn. Pearson Prentice Hall (2008)
2. Goldberg, Y.: A primer on neural network models for natural language processing. J. Artif. Intell. Res. **57**(2016), 345–420 (2016)
3. Saeed, K., Nammous, M.K.: A speech-and-speaker identification system: feature extraction, description, and classification of speech-signal image. IEEE Trans. Ind. Electron. **54**(2), 887–897 (2007)
4. Nammous, M.K., Szczepanski, A., Saeed, K.: An exploratory research on text-independent speaker recognition. In: HAIS, Part 1, pp. 412–419 (2011)
5. Ahmed, H., Elaraby, M.S., Moussa, A.M., Abdallah, M., Abdou, S.M., Rashwan, M.: An unsupervised speaker clustering technique based on SOM and I-vectors for speech recognition systems. In: The Third Arabic Natural Language Processing Workshop, EACL, Valencia, Spain (2017)
6. Sarria-Paja, M., Falk, T.H.: Variants of mel-frequency cepstral coefficients for improved whispered speech speaker verification in mismatched conditions. In: 25th European Signal Processing Conference (EUSIPCO) (2017)
7. Lopez-Otero, P., Docio-Fernandez, L., Garcia-Mateo, C.: I-vectors for continuous emotion recognition. Training **45**, 50 (2014)
8. Bahari, M.H., Mclaren, M., Van Hamme, H., Van Leeuwen, D.A.: Speaker age estimation using I-vectors. Eng. Appl. Artif. Intell. **34**, 99–108 (2014)
9. Motlicek, P., Dey, S., Madikeri, S., Burget, L.: Employment of subspace gaussian mixture models in speaker recognition. In: 2015 IEEE International Conference on Acoustics, Speech and Signal Processing (ICASSP), South Brisbane, QLD, pp. 4445–4449 (2015)
10. Saeed, K.: Carathéodory–Toeplitz based mathematical methods and their algorithmic applications in biometric image processing. Appl. Numer. Math. **75**, 2–21 (2014)
11. Specht, D.F.: Probabilistic neural networks and the polynomial adaline as complementary techniques for classification. IEEE Trans. Neural Netw. **1**, 11–121 (1990)
12. Low, R., Togneri, R.: Speech recognition using the probabilistic neural network. In: Proceedings of ICSLP98 (1998)
13. Phan, H., Koch, P., Katzberg, F., Maass, M., Mazur, R., Mertins, A.: Audio scene classification with deep recurrent neural networks (2017). arXiv:1703.04770
14. Qawaqneh, Z., Mallouh, A.A., Barkana, B.D.: Deep neural network framework and transformed MFCCs for speaker's age and gender classification. Knowl. Based Syst. **115**, 5–14 (2017)

15. Becerra, A., de la Rosa, J.I., González, E.: Speech recognition in a dialog system: from conventional to deep processing. In: Multimedia Tools and Applications, pp. 1–37. Springer (2017)
16. López Moreno, I.: Deep neural network architectures for large-scale, robust and small-footprint speaker and language recognition. Ph.D. thesis. Universidad Politécnica de Madrid (2017)
17. Bell, P., Gales, M., Hain, T., Kilgour, J., Lanchantin, P., Liu, X., McParland, A., Renals, S., Saz, O., Wester, M., Woodland, P.: The MGB challenge: evaluating multi-genre broadcast media recognition. In: IEEE Workshop on Automatic Speech Recognition and Understanding, pp. 687–693. IEEE (2015)
18. Feng, L., Hansen, L.K.: A new database for speaker recognition. Technical report (2005)
19. McLaren, M., Ferrer, L., Castán, D., Lawson, A.: The speakers in the wild (SITW) speaker recognition database. In: INTERSPEECH, vol. 2016, pp. 818–822 (2016)
20. Woo, R.H., Park, A., Hazen, T.J.: The MIT mobile device speaker verification corpus: data collection and preliminary experiments. In: The Speaker and Language Recognition Workshop (2006)
21. Godfrey, J.J., Holliman, E.C., McDaniel, J.: Switchboard: telephone speech corpus for research and development. In: Proceedings of the IEEE International Conference on Acoustics, Speech and Signal Processing, vol. 1, pp. 517–520. IEEE (1992)
22. Greenberg, C.S.: The NIST year 2012 speaker recognition evaluation plan. NIST, Technical report (2012)
23. Poignant, J., Besacier, L., Quénot, G.: Unsupervised speaker identification in TV broadcast based on written names. IEEE/ACM Trans. Audio Speech Lang. Process. 23(1) (2015)
24. Nagraniy, A., Chungy, J.S., Zisserman, A.: VoxCeleb: a large-scale speaker identification dataset. In: INTERSPEECH (2017)
25. Nammous M., Saeed K.: Voice-print and text-independent speaker identification. In: International Conference on Electrical Engineering Design and Technologies—ICEEDT'07, 1 Jan 2007. International Conference on Electrical Engineering Design and Technologies—ICEEDT'08, Tunisia (2007)
26. Bishop, C.M.: Pattern Recognition and Machine Learning. Springer, New York, NY (2006)
27. Kusy, M., Zajdel, R.: Probabilistic neural network training procedure based on Q(0)-learning algorithm in medical data classification. Appl. Intell. 41, 837–854 (2014)
28. Specht, D.F.: Probabilistic neural networks. Neural Netw. 3(1), 109–118 (1990)
29. Lewicki, P., Hill, T.: Statistics: Methods and Applications: a Comprehensive Reference for Science, Industry, and Data Mining. StatSoft Inc, Tulsa, OK (2006)
30. Hochreiter, S., Schmidhuber, J.: Long short-term memory. Neural Comput. 9(8), 1735–1780 (1997)
31. Zhou, P., Qi, Z., Zheng, S., Xu, J., Bao, H., Xu, B.: Text classification improved by integrating bidirectional LSTM with two-dimensional max pooling. In: COLING 2016, pp. 3485–3495 (2016)
32. Lu, Y., Lu, C., Tang, C.-K.: Online video object detection using association LSTM. In: The IEEE International Conference on Computer Vision (ICCV), pp. 2344–2352 (2017)
33. Akopyan, M., Khashba, E.: Large-scale YouTube-8M video understanding with deep neural networks (2017). arXiv:1706.04488
34. Xu, J., Chen, D., Qiu, X., Huang, X.: Cached long short-term memory neural networks for document-level sentiment classification. In: EMNLP 2016, pp. 1660–1669 (2016)
35. Lu, L., Renals, S.: Small-footprint highway deep neural networks for speech recognition. IEEE/ACM Trans. Audio Speech Lang. Process. 25(7), 1502–1511 (2017)
36. Chen, J., Wang, D.L.: Long short-term memory for speaker generalization in supervised speech separation. In: INTERSPEECH, pp. 3314–3318 (2016)
37. Saeed, K., Adamski, M., Bhattasali, T., Nammous, M.K., Panasiuk, P., Rybnik, M., Shaikh, S.H.: New Directions in Behavioral Biometrics. CRC Press (2016)

Part IV
Image Processing

Heterogeneous Face Matching Using Robust Binary Pattern of Local Quotient: RBPLQ

Hiranmoy Roy and Debotosh Bhattacharjee

Abstract This paper proposes a robust binary scheme for representing and matching near-infrared (NIR)–visible (VIS) and sketch–photo heterogeneous faces. It is termed as robust binary pattern of local quotient (*RBPLQ*). *RBPLQ* provides illumination-invariant and noise-resistant features in coarse level. At first, a local quotient (*LQ*) is extracted for representing illumination-invariant image. Then, a robust local binary feature is proposed to capture the variations of *LQ*. The proposed technique is applied to different benchmark databases of NIR-VIS and sketch–photo images. Recognition accuracy of 60.72% is achieved in the NIR-VIS database. In the CUFSF database, which is a viewed sketch–photo database, the recognition accuracy of 96.24% is achieved. Extended Yale B database is also tested for verifying the illumination-invariant property of *RBPLQ*, and it achieved recognition accuracy of 94.20%. Finally, *RBPLQ* also provides good performance in the case of noisy situations.

Keywords Local quotient image · Modality-invariant feature
Illumination-reflectance model · Robust binary pattern
Illumination-invariant feature · Heterogeneous face recognition

1 Introduction

One of the most essential and classical computer vision problems is the face recognition. Variety of methodologies have been reported to solve the practical and real-life face recognition problems. Due to the recent advancement in technology, a variety of face capturing equipments have been developed depending on the real-life

H. Roy (✉)
Department of Information Technology, RCC Institute of Information Technology,
Canal South Road, Beliaghata, Kolkata 700015, India
e-mail: hiranmoy.roy@rcciit.org

D. Bhattacharjee
Department of Computer Science and Engineering, Jadavpur University,
Kolkata 700032, India
e-mail: debotoshb@hotmail.com

© Springer Nature Singapore Pte Ltd. 2019
R. Chaki et al. (eds.), *Advanced Computing and Systems for Security*,
Advances in Intelligent Systems and Computing 883,
https://doi.org/10.1007/978-981-13-3702-4_10

requirements. At night, face images are easily captured using newly developed near-infrared (NIR) cameras. The same NIR cameras also used in face recognition with varying illuminations [1]. Another type of thermal-infrared (TIR) cameras is very useful to sense the heat radiated from the body for authentication as well as liveness detection. Whereas in forensic, face sketches are very useful for detecting the criminals. Sometimes, it may happen that either there is no available camera or the camera is there but it has captured images whose quality is very poor. In such situations, the only solution is the face sketches drawn from the verbal information presented by an eyewitness. Therefore, the accumulation of face images for real-world applications is from different environments. For obvious reason, these face images are not from the same modality and called heterogeneous face images. The related real-world face recognition problem is named as heterogeneous face recognition (HFR) [2]. Therefore, the conventional homogeneous face recognition systems are not always suitable to use in these HFR situations.

Recently, the problem of heterogeneous face recognition has gained increasing popularity. In the literature, many different methodologies have been developed to solve the HFR problem. These solutions are easily categorized into the following three broad categories:

- **Synthesis-based methods**: The main idea of this category is to transfer one modality faces into another by using synthesis techniques. The transformed photo or sketch is called a pseudo-photo or pseudo-sketch, respectively. Then some classification technique is used for recognition. Tang and Wang [3] introduced sketch-photosynthesis method using eigen transformation. Chen et al. [4] used the same eigen transformation for NIR-VIS synthesis. Gao et al. [5] embedded a hidden Markov model (HMM) along with an ensemble technique for photo–sketch synthesis. A Markov random field (MRF) model for facial patch-based synthesis from sketches into photos is proposed by Wang and Tang [6]. The same MRF model was also adopted by Li et al. [7] for synthesizing TIR images into VIS images. A synthesis technique by transductive learning (TFSPS) was developed by Wang et al. [8]. Gao et al. [9] presented sketch–photo and photo–sketch synthesis using sparse representation. Another interesting synthesis technique was presented by Wang et al. [10] using sparse feature selection (SFS) and support vector regression. Recently, a multiple representation-based synthesis for face sketch–photo was proposed by Peng et al. [11]. Since sketch or photo faces are converted into same modality, available homogeneous face recognition classifiers are used. The main drawback of the methods in this category is the time-consuming synthesis step. Again, the synthesis process is not generalized. Therefore, for different scenarios such as NIR-VIS synthesis, and photo–sketch synthesis, different learning mechanism is required.

- **Subspace learning-based methods**: In this category, projected subspace learning for different modality face images are used to minimize the modality gap. A subspace learning using common discriminant feature extraction (CDFE) for recognizing face sketch–photo images was introduced by Lin and Tang [12]. A subspace learning method based on partial least square (PLS) was proposed by

Sharma and Jacobs [13]. Yi et al. [14] developed a canonical correlation method for NIR-VIS matching. Another subspace learning for NIR-VIS matching, called coupled spectral regression (CSR), was proposed by Lei and Li [15]. Mignon and Jurie [16] presented a metric learning based on cross-modality (CMML) for heterogeneous faces. A multi-view discriminant analysis (MvDA) for single discriminant common subspace was developed by Kan et al. [17]. Another coupled discriminant analysis for HFR was proposed by Lei et al. [18]. Subspace learning easily reduces the modality gaps present between images of different modalities. The main drawback of the methods in this category is the information loss due to projection of images. Similar to the first category, the subspace learning method is time-consuming.

- **Modality-invariant feature-based methods**: In this category, some modality-invariant feature representation methods are used to represent the images of different modalities. Liao et al. [19] represented NIR and VIS images in some modality-invariant features using multi-block local binary pattern (MB-LBP) and difference of Gaussian (DoG) filter. The multi-scale local binary pattern (MLBP) and scale-invariant feature transform (SIFT) features were employed by Klare et al. [20] for forensic sketch recognition. Zhang et al. [21] represented the sketch and photo faces in a form of encoding tree called coupled information-theoretic encoding (CITE). For semi-forensic sketches, a multi-scale circular Weber's local descriptor (MCWLD) was used by Bhatt et al. [22]. Again, on MLBP features a random subspace, called kernel prototype (KP-RS), was proposed by Klare and Jain [23]. Using Log-DoG filter, histogram of oriented gradient (HOG), and LBP features, Zhu et al. [24] developed a transductive learning (THFM) technique. The same HOG and MLBP features were used for canonical correlation analysis (MCCA) by Gong et al.[25]. Roy and Bhattacharjee [26–31] mostly emphasized on modality-invariant features like edges and textures. Since edges and textures both are illumination sensitive, Roy et al. tried to implement some illumination-invariant edge and texture capturing features. Peng et al. [32] developed a graph-based feature representation for HFR called G-HFR. Handcrafted features extraction methods are directly applied, which means more discriminative local features are available. Again, methods are more time saving than other two categories. This category faces only one major challenge and it is to recognize or search modality-invariant features.

The advantages of modality-invariant feature are the main motivation to propose a new feature for heterogeneous face matching. Identification of modality-invariant facial features for heterogeneous faces is the aim of the proposed method. From psychological studies [33] and using our visual inspection, we understand that edge is one of the key modality-invariant facial features. We used the same motivation in our previous works [26–28, 34] also. Maximum edges are found in almost all the facial components like eyes and nose. These facial components also belong to high-frequency information of an image. At the same time, every face has unique texture feature, which is also important for matching faces. Again, these two features, i.e., edge and texture on a face image, are very much illumination sensitive. Therefore, not only an illumination-invariant domain is required, but also the capability of capturing high-frequency information is essential. If we closely inspect the drawing of a sketch, then we will easily see that the artist provides more interest toward texture and edge information. Furthermore, the NIR images are developed from the high-frequency features. Therefore, we can conclude that we sensibly selected the most important two modality-invariant features and they are edge and texture. We propose a local quotient (LQ) image representation to make it illumination insensitive. LQ provides an image with high-frequency features. It is also very important to extract local micro-level texture feature. Motivated by the local binary pattern (LBP) [35] and LBP-like features, which give superior results in the texture as well as face recognition, we propose a new binary pattern. Although, in our previous works [26, 34], we use different techniques to capture local texture, those methods are not robust in nature, i.e., noise sensitive. Therefore, we propose one novel robust binary pattern (RBP) this time. RBP measures the binary relation between pixels with a modified threshold value. RBP is more noise resistive than LBP. At last, the idea of combining LQ and RBP develops the proposed modality-invariant feature for HFR. We name it a robust binary pattern of local quotient ($RBPLQ$). Different HFR databases are tested to show the superior result of the proposed technique.

The major contributions are:

1. LQ is developed for representing illumination-invariant domain.
2. RBP is proposed to measure the relation between pixels in a local window with more noise resistance power than LBP.
3. $RBPLQ$ is proposed to collect the edge patterns and local textures of the key facial components.

The organization of the paper is as follows: The detail description of the proposed $RBPLQ$ is given in Sect. 2. In Sect. 3, the experimental results and comparisons are presented, and finally, Sect. 4 concludes the paper.

2   Proposed Work

This section starts with an introduction of the technique of modality-invariant feature extraction for HFR. At first, a detailed idea about the image representation LQ for illumination-invariant domain is provided. Finally, we conclude with a detail explanation of the proposed $RBPLQ$ feature.

2.1   Image Representation Using Local Pixels' Quotient

Any face recognition system suffers from some external effects, among them, illumination variations are one of the major effects. Illumination variations heavily increase the intra-class variation between faces. Again, our selected modality-invariant feature, i.e., edges are also illumination sensitive. Therefore, it is really essential to represent the image in illumination-invariant domain, so that better edges can be extracted. The illumination-reflectance model (IRM) [36, 37] explains that each image pixel of a gray face image $I(x, y)$ is represented as the product of the reflectance part $R(x, y)$ and the illumination part $L(x, y)$, as shown in the following equation:

$$I(x, y) = R(x, y) \times L(x, y) \tag{1}$$

Here, the R part holds the feature representation of the key facial edges and points. The L part reflects the amount of light that falls on the face. Now, if we are able to elimination the L part from a face image, the R part will be able to represent most of the key facial features like edges. It means that the R part itself is the most costly feature for representing the image in modality-invariant domain. Moreover, the R part holds the high-frequency representation of an image; on the other hand, L part holds the low-frequency representation. From the literature [27, 38], we have found the widely recognized assumption, i.e., over a local 3×3 neighborhood L part approximately remains constant.

In the literature, varieties of techniques have been developed to decrease the effect of variations of illumination. Those methods are easily classified into two categories based on mathematical operations: subtraction and division. In [39, 40], authors applied subtraction operation, whereas in [27, 38, 41] authors applied division operation. The only problem of subtraction-based methods is to apply the logarithmic operation to get additive relation from the multiplicative IRM relation. Since the division operation is better than the subtraction operation (as shown in [27]) due to no need for logarithmic conversion, it is better to apply division-based approach to reduce the L part.

Two different local average values are measured from the 3×3 image patch. One average (Avg_{dir}) is measured by considering the pixels, which are directly connected with the central pixel (as shown in Fig. 1a), and another average Avg_{diag} is measured by considering the diagonally connected pixels with the central pixel (as shown in Fig. 1b).

The measuring of the two averages is done using the following equations:

Fig. 1 **a** Directly connected pixels with the central pixel in a 3×3 window, **b** diagonally connected pixels with the central pixel in a 3×3 window

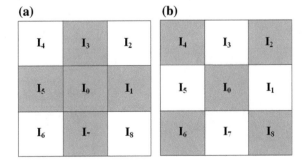

$$Avg_{dir} = \frac{1}{5} \left(I_0 + \sum_{i=0}^{3} I_{(2i+1)} \right) \tag{2}$$

$$Avg_{diag} = \frac{1}{5} \left(I_0 + \sum_{i=0}^{3} I_{(2i+2)} \right) \tag{3}$$

Then, the local pixel quotient (LQ) is measured in the following equation:

$$LQ = \frac{Avg_{dir}}{Avg_{diag}} \tag{4}$$

The division operation creates one major problem in LQ calculation. If all the pixels in the diagonal position of the central pixel in a 3×3 window are with an intensity value zero, then Avg_{diag} and LQ become zero and infinity, respectively. This major problem is solved by translating the grayscale interval from [0, 255] to [1, 256]. It is done by adding one with each pixel value. This translation makes the image as well as LQ approximately finite.

Now, Eq. 4 can be written as follows

$$LQ = \frac{\frac{1}{5} \left(I_0 + \sum_{i=0}^{3} I_{(2i+1)} \right)}{\frac{1}{5} \left(I_0 + \sum_{i=0}^{3} I_{(2i+2)} \right)} \tag{5}$$

According to the IRM (Eq. 1), Eq. 5 is modified to:

$$LQ = \frac{(R_0 \times L_0) + \sum_{i=0}^{3} \left(R_{(2i+1)} \times L_{(2i+1)} \right)}{(R_0 \times L_0) + \sum_{i=0}^{3} \left(R_{(2i+2)} \times L_{(2i+2)} \right)} \tag{6}$$

Since L part is assumed to be constant over a local 3×3 neighborhood and let the value is L_c, Eq. 6 is modified as follows

$$LQ = \frac{(R_0 \times L_c) + \sum_{i=0}^{3} \left(R_{(2i+1)} \times L_c \right)}{(R_0 \times L_c) + \sum_{i=0}^{3} \left(R_{(2i+2)} \times L_c \right)}$$

$$LQ = \frac{\left(R_0 + \sum_{i=0}^{3} R_{(2i+1)} \right) \times L_c}{\left(R_0 + \sum_{i=0}^{3} R_{(2i+2)} \right) \times L_c}$$

$$\Rightarrow LQ = \frac{\left(R_0 + \sum_{i=0}^{3} R_{(2i+1)} \right)}{\left(R_0 + \sum_{i=0}^{3} R_{(2i+2)} \right)} \tag{7}$$

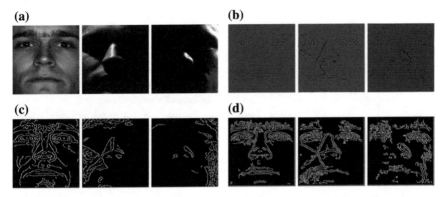

Fig. 2 **a** Few images with varying illuminations, **b** the corresponding proposed LQ images, **c** the edge detected images using canny method of (**a**) images, **d** the edge detected images using canny method of (**b**) LQ images

Now, if we look at the final measure of LQ, then we can see that the L part is vanished and it has only the R part. We already know that R part represents the important modality-invariant features, which is the required essential feature for HFR. Since in LQ there is no L part, we can say that LQ represents the illumination-invariant domain of the image.

In the literature, many quotient image representation techniques [41, 42] have been proposed for solving illumination effect. These quotient image techniques applied normal division operation between the original image and a blurred version of the original image. In our approach, we have also applied a division operation, but it is robust in nature than the methods, which are mentioned above. Since a local averages are measured to get Avg_{diag} and Avg_{dir}, it acts as an average filtering. This by default average filtering makes our method more noise resistant than others. The experimental results also suggest the same thing. Thus, it is worth to say that our LQ image representation is illumination-invariant as well as noise resistant. Few sample images from Extended Yale B database with varying illuminations and their corresponding LQ images are shown in Fig. 2.

In Fig. 2d, the proposed LQ images show better edges than normal quotient method after applying canny edge detection algorithm.

2.2 Robust Binary Pattern Generation

Local binary pattern (LBP) [35] has emerged as one of the best local features. It has been successfully applied in many applications related to image processing and pattern recognition. The binary relations between the central pixel against its neighboring pixels represent the LBP feature. When neighboring pixel is greater than or equal to center pixel, then the relation is represented as binary value '1' otherwise '0'. LBP mainly uses the central pixel's value as the threshold value to represent the

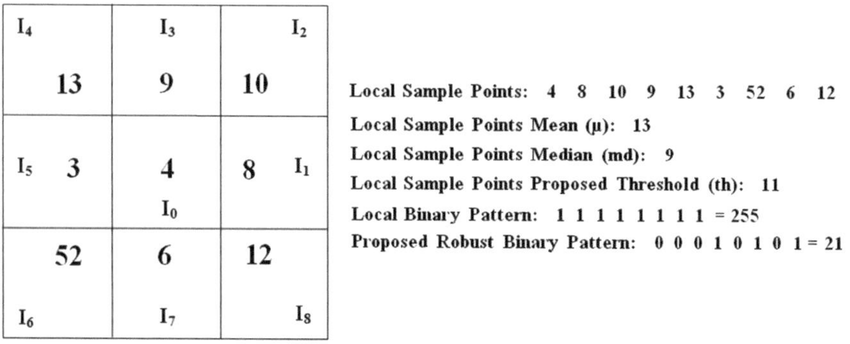

I_4	I_3	I_2
13	9	10
I_5 3	4	8 I_1
	I_0	
52	6	12
I_6	I_7	I_8

Local Sample Points: 4 8 10 9 13 3 52 6 12
Local Sample Points Mean (µ): 13
Local Sample Points Median (md): 9
Local Sample Points Proposed Threshold (th): 11
Local Binary Pattern: 1 1 1 1 1 1 1 1 = 255
Proposed Robust Binary Pattern: 0 0 0 1 0 1 0 1 = 21

Fig. 3 Example to obtain proposed *RBP* from a 3 × 3 neighborhood and comparison with LBP

binary pattern. Although LBP represents local features in microstructures, it cannot properly deal with noise and different lighting conditions. To solve these problems, we proposed a robust threshold, which is measured by averaging the local mean (μ) and median (md). Figure 3 shows the 3 × 3 image patch used in our experiment and how the proposed threshold (th) is measured.

We have considered that the image pixels are in a square window, and they are shown in Fig. 3. Then, the mean of all those pixels are measured as follows:

$$\mu = \frac{1}{P} \sum_{i=0}^{P} I_i \tag{8}$$

where P represents the number of pixels in a square window. The value of P becomes 9 when a 3 × 3 window is considered. The value of the median (md) is calculated by sorting all the $P = 9$ pixels and choosing the $\lceil \frac{P}{2} \rceil = 5$th pixel. Finally, the proposed robust threshold is calculated as follows:

$$th = \frac{\mu + md}{2} \tag{9}$$

Now, the measured 'th' is used to calculate the binary relations using the following equation(for $P=8$):

$$RBP = \sum_{i=1}^{P} 2^{(i-1)} \times f(I_i - th)$$

$$f(a) = \begin{cases} 1, & if a \geq 0 \\ 0, & otherwise \end{cases} \tag{10}$$

To measure the texture pattern present in LQ, the proposed *RBP* is applied. Finally, the local texture pattern represents our proposed *RBPLQ*.

(a)

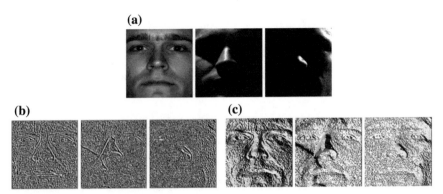

(b)   **(c)**

Fig. 4 **a** Few images with varying illuminations, **b** the corresponding proposed *RBPLQ* images, **c** the corresponding LBP images

(a)   **(b)**   **(c)**   **(d)**

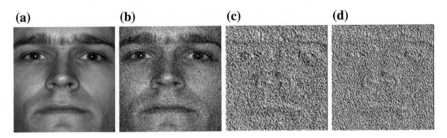

Fig. 5 **a** Original image, **b** corresponding Gaussian noise (mean = 0, and standard deviation = 0.01) added image, **c** corresponding proposed *RBPLQ* image, **d** corresponding LBP image

The proposed robust binary pattern not only gives the better noise resistant feature, but also gives an edge-preserving texture feature. In Fig. 4, the different LBP images and *RBPLQ* images show the better edge-preserving property of *RBPLQ* than LBP. Figure 5 shows the better noise resistant property than LBP. It is visible from the figure that *RBPLQ* preserves far better facial features than normal LBP.

2.3 Similarity Measure

In any LBP-based technique, the similarity between a query image and gallery image is measured by block-wise histogram matching. At first, the binary image is partitioned into non-overlapping square blocks. Here we have used non-overlapping square blocks of dimension $w_b \times w_b$ ($w_b = 6, 8, 10, 12$) to partition the *RBPLQ* image. Histogram of each partitioned blocks is measured and concatenated. The concatenated histogram represents the feature vector for the face. Simple nearest neighbor (NN) classifier with a distance-based similarity measure, called histogram intersection, is used. The similarity score of *RBPLQ* $S(I_q, I_G)$ between a query image (I_q) and gallery image (I_G) is measured as follows:

$$S(I_q, I_G) = - \sum_{i,j} \min \left(H_q^{i,j}, H_G^{i,j} \right) \tag{11}$$

where $H_q^{i,j}$ and $H_G^{i,j}$ are the concatenated histograms of the query and gallery images, respectively. Again, (i, j) represents the jth bin of ith block.

3 Experimental Results

This section explains the performance results of *RBPLQ* on different databases. One benchmark database for testing illumination-invariant property and two benchmark databases for testing modality-invariant property of *RBPLQ* are used here. Extended Yale B Face database [43, 44] is a well-known benchmark database for testing illumination-invariant property. CASIA NIR-VIS 2.0 Face Database [45] is used for NIR-VIS face recognition. The CUHK Face Sketch FERET Database (CUFSF) [21] is used for face sketch–photo recognition. To test the noise resistant property of the proposed *RBPLQ*, every query image is added with Gaussian noise (mean = 0 and Sd = 0.01) and matched with gallery images.

At first, the illumination-invariant property of the proposed *RBPLQ* is tested. The rank-1 recognition accuracy of the proposed technique is compared with several state-of-the-art techniques: LBP, TVQI [41], HF+HQ [40], Gradient-face [38], MSLDE [39]. These methods are implemented and tuned as per their published papers.

To measure the performance of the proposed *RBPLQ* for modality-invariant NIR-VIS HFR, it is compared with several state-of-the-art techniques: MvDA and G-HFR. Again, *RBPLQ* is compared with techniques: CITE [21], KP-RS [23], MCCA [25], PLS [13], TFSPS [8], G-HFR [32], and MvDA [17] on viewed sketch database. Experimental setups and recognition accuracies of these techniques are taken from their published papers.

3.1 Recognition Results on Extended Yale B Database

This database is a well-accepted benchmark database of 2432 face images of 38 different subjects with 64 varying illumination conditions. All face images are resized to 120×120 pixels. We have separated the database into five different subsets based on the illumination angles. Table 1 gives the description of the different subsets. Figure 6 shows one sample face from the database under varying illumination conditions along with the corresponding *RBPLQ* faces. Neutral light images from Subset 1 was set as the gallery. Rest images from Subset 1 to Subset 5 became the query images. Rank-1 accuracy results are shown on the individual subset and after averaging over all subsets in Table 2. Proposed *RBPLQ* performed better than other methods on not only illumination situations but also in a noisy situation.

Table 1 Different subsets of Extended Yale B database according to their illumination angle

Subset names	Angle of illumination	Number of images per subject
S1	0°–12°	7
S2	13°–25°	12
S3	26°–50°	12
S4	51°–77°	14
S5	78° and above	19

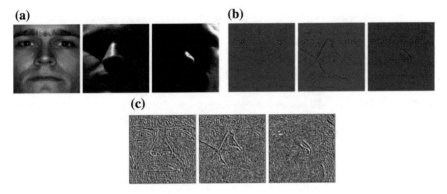

Fig. 6 **a** Sample illumination variation images from Extended Yale B Face database, **b** corresponding *LQ* images, **c** corresponding *RBPLQ* images

Table 2 Recognition results for different methods on different subsets of Extended Yale B database of 38 subjects

State-of-the-art methods	Recognition accuracy (%)						
	S1	S2	S3	S4	S5	Avg	Avg (Noise)
LBP	84.81	89.91	80.04	72.18	74.99	80.39	35.12
TVQI	92.28	95.20	90.27	81.39	84.32	88.69	57.78
HF+HQ	94.81	98.76	93.18	82.90	84.43	90.82	58.31
Gradient-face	94.74	**100**	83.33	75.94	84.65	87.73	67.36
MSLDE	96.61	**100**	93.20	86.66	89.33	93.16	60.09
RBPLQ	**96.93**	**100**	**94.01**	**90.09**	**89.97**	**94.2**	**70.09**

3.2 Recognition Results on CASIA NIR-VIS 2.0 Database

CASIA NIR-VIS 2.0 database consists of 17,580 near-infrared and visible light images of 725 subjects. The images also have some variations in pose, age, and expression. We have tested the near-infrared images as probe images and visible images as a gallery. We manually cropped all the front faces and maintained almost the same eye levels. Then, those cropped images are resized to 120 × 120 pixels. Standard evaluation protocols as given in [46] are followed here.

Table 3 Recognition results for different methods on CASIA NIR-VIS 2.0 database of 725 subjects

State-of-the-art methods	Training samples (number of subjects)	Testing samples (number of subjects)	Recognition accuracy (%)	Recognition accuracy (%) (Noise)
MvDA	517	208	41.60	10.24
G-HFR	517	208	54.90	15.67
RBPLQ	0	208	**60.72**	**20.09**

(a)   (b)

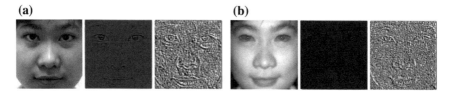

Fig. 7 Normal VIS and NIR images and their corresponding *RBPLQ* images from CASIA NIR-VIS 2.0 database, **a** row-wise: VIS image, corresponding *LQ* image, and corresponding *RBPLQ* image, **b** row-wise: NIR image, corresponding *LQ* image, and corresponding *RBPLQ* image

The recognition result is compared against two state-of-the-art methods (MvDA, and G-HFR). In Table 3, rank-1 recognition accuracy of the proposed method compared with MvDA and G-HFR is given. The proposed method achieved 60.72% accuracy and which is better than other two methods. Recognition result of MvDA and G-HFR is taken from their corresponding published papers. Sample image is shown in Fig. 7 to represent the application of the proposed *RBPLQ* feature on NIR and VIS images from CASIA NIR-VIS 2.0 database.

3.3   Recognition Results on Viewed Sketch Databases

In this experiment, we used A CUHK Face Sketch FERET (CUFSF) database. This is a well-known benchmark database for viewed sketch versus photo recognition. CUFSF database has 1194 different subjects from the FERET database. An artist drew the sketches with shape exaggeration and light variations by viewing the photos of each person. All the images are manually cropped by maintaining the same eye levels for the frontal faces. Then, those cropped images are resized to a dimension of 120×120 pixels. Recognition results of several existing state-of-the-art techniques (CITE, KP-RS, MCCA, PLS, TFSPS, G-HFR, and MvDA) are compared. Experimental results of other techniques are collected from their published papers. The proposed *RBPLQ* achieved 96.24% rank-1 accuracy. In Table 4, comparison result is given. Proposed method outperforms other state-of-the-art techniques not only in a normal situation but also in a noisy situation. Sample image is shown in Fig. 8 to represent the application of the proposed *RBPLQ* feature on face photo and sketch images from the CUFSF database.

Table 4 Recognition results for different methods on CUFSF database

State-of-the-art methods	Training samples (number of subjects)	Testing samples (number of subjects)	Recognition rate (%)	Recognition rate (%) (Noise)
CITE	500	694	89.54	61.98
KP-RS	500	694	83.95	58.20
MCCA	300	894	92.17	67.31
PLS	300	894	51.00	31.56
TFSPS	300	894	72.62	50.11
G-HFR	500	694	96.04	71.38
MvDA	500	694	55.50	19.37
RBPLQ	0	1194	**96.24**	**76.09**

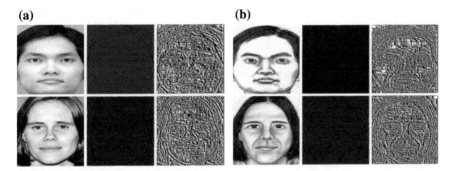

(a)   **(b)**

Fig. 8 Normal photo and sketch images and their corresponding *RBPLQ* images from CUFSF database. **a** Row-wise: photo images, corresponding *LQ* images, and corresponding *RBPLQ* images, **b** row-wise: sketch images, corresponding *LQ* images, and corresponding *RBPLQ* images

4 Conclusion

RBPLQ is a novel feature mainly for illumination-invariant and modality-invariant face recognition. It is a mixture or hybrid kind feature combination of *LQ* and *RBP*. The combination feature really increases the recognition performance of heterogeneous faces. We showed theoretically that the proposed *LQ* is not only illumination insensitive but also edge-preserving representation of an image. Local patterns of *LQ* are captured using a robust binary pattern, called (*RBP*). In noisy situations, almost all the LBP-like features failed to give high accuracy. Therefore, the proposed robust binary pattern is very much useful in noisy situations also.

Experimental studies on different situations like variations of illumination, viewed sketch versus photo matching, NIR versus VIS face matching, and the presence of noise *RBPLQ* gives better performance than other comparable techniques.

In the future, we intend to apply *RBPLQ* in different other applications of image processing, mainly in the field of texture analysis. Further, in the era of deep learning a combination of handcrafted (*RBPLQ*) and deep learning features to improve the recognition accuracies can be thought.

References

1. Li, S., Chu, R., Liao, S., Zhang, L.: Illumination invariant face recognition using NIR images. IEEE Trans. Pattern Anal. Mach. Intell. **29**(4), 627–639 (2007)
2. Li, S.: Encyclopaedia of Biometrics. Springer (2009)
3. Tang, X., Wang, X.: Face sketch recognition. IEEE Trans. Circuits Syst. Video Technol. **14**(1), 50–57 (2004)
4. Chen, J., Yi, D., Yang, J., Zhao, G., Li, S., Pietikainen, M.: Learning mappings for face synthesis from near infrared to visual light images. In: Proceedings of IEEE International Conference on Computer Vision and Pattern Recognition, pp. 156–163 (2009)
5. Gao, X., Zhong, J., Li, J., Tian, C.: Face sketch synthesis algorithm on E-HMM and selective ensemble. IEEE Trans. Circuits Syst. Video Technol. **18**(4), 487–496 (2008)
6. Wang, X., Tang, X.: Face photo-sketch synthesis and recognition. IEEE Trans. Pattern Anal. Mach. Intell. **31**(1), 1955–1967 (2009)
7. Li, J., Hao, P., Zhang, C., Dou, M.: Hallucinating faces from thermal infrared images. In: Proceedings IEEE International Conference on Image Processing, pp. 465–468 (2008)
8. Wang, N., Tao, D., Gao, X., Li, X., Li, J.: Transductive face sketch-photo synthesis. IEEE Trans. Neural Netw. **24**(9), 1364–1376 (2013)
9. Gao, X., Wang, N., Tao, D., Li, X.: Face sketchphoto synthesis and retrieval using sparse representation. IEEE Trans. Circuits Syst. Video Technol. **22**(8), 1213–1226 (2012)
10. Wang, N., Li, J., Tao, D., Li, X., Gao, X.: Heterogeneous image transformation. Elsevier J. Pattern Recogn. Lett. **34**, 77–84 (2013)
11. Peng, C., Gao, X., Wang, N., Tao, D., Li, X., Li, J.: Multiple representation-based face sketch-photo synthesis. IEEE Trans. Neural Netw. **xxx**, 1–13 (2016)
12. Lin, D., Tang, X.: Inter-modality face recognition. In: Proceedings of European Conference on Computer Vision, pp. 13–26 (2006)
13. Sharma, A., Jacobs, D.: Bypassing synthesis: PLS for face recognition with pose, low-resolution and sketch. In: Proceedings of IEEE International Conference on Computer Vision and Pattern Recognition, pp. 593–600 (2011)
14. Yi, D., Liu, R., Chu, R., Lei, Z., Li, S.: Face matching between near infrared and visible light images. In: Proceedings of International Conference on Biometrics, pp. 523–530 (2007)
15. Lei, Z., Li, S.: Coupled spectral regression for matching heterogeneous faces. In: Proceedings of IEEE International Conference on Computer Vision and Pattern Recognition, pp. 1123–1128 (2009)
16. Mignon, A., Jurie, F.: CMML: a new metric learning approach for cross modal matching. In: Proceedings of Asian Conference on Computer Vision, pp. 1–14 (2012)
17. Kan, M., Shan, S., Zhang, H., Lao, S., Chen, X.: Multi-view discriminant analysis. IEEE Trans. Pattern Anal. Mach. Intell. **38**(1), 188–194 (2016)
18. Lei, Z., Liao, S., Jain, A.K., Li, S.Z.: Coupled discriminant analysis for heterogeneous face recognition. IEEE Trans. Inf. Forensics Secur. **7**(6), 1707–1716 (2012)
19. Liao, S., Yi, D., Lei, Z., Qin, R., Li, S.: Heterogeneous face recognition from local structure of normalized appearance shared representation learning for heterogeneous face recognition. In: Proceedings of IAPR International Conference on Biometrics (2009)
20. Klare, B.F., Li, Z., Jain, A.K.: Matching forensic sketches to mug shot photos. IEEE Trans. Pattern Anal. Mach. Intell. **33**(3), 639–646 (2011)
21. Zhang, W., Wang, X., Tang, X.: Coupled information-theoretic encoding for face photo-sketch recognition. In Proceedings of IEEE International Conference on Computer Vision and Pattern Recognition, pp. 513–520 (2011)

22. Bhatt, H.S., Bharadwaj, S., Singh, R., Vatsa, M.: Memetically optimized MCWLD for matching sketches with digital face images. IEEE Trans. Inf. Forensics Secur. **7**(5), 1522–1535 (2012)
23. Klare, B.F., Jain, A.K.: Heterogeneous face recognition using kernel prototype similarities. IEEE Trans. Pattern Anal. Mach. Intell. **35**(6), 1410–1422 (2013)
24. Zhu, J., Zheng, W., Lai, J., Li, S.: Matching NIR face to VIS face using transduction. IEEE Trans. Inf. Forensics Secur. **9**(3), 501–514 (2014)
25. Gong, D., Li, Z., Liu, J., Qiao, Y.: Multi-feature canonical correlation analysis for face photo-sketch image retrieval. In: Proceedings of ACM International Conference on Multimedia, pp. 617–620 (2013)
26. Roy, H., Bhattacharjee, D.: Heterogeneous face matching using geometric edge-texture feature (GETF) and multiple fuzzy-classifier system. Elsevier J. Appl. Soft Comput. **46**, 967–979 (2016)
27. Roy, H., Bhattacharjee, D.: Local-gravity-face (LG-face) for illumination-invariant and heterogeneous face recognition. IEEE Trans. Inf. Forensics Secur. **11**(7), 1412–1424 (2016)
28. Roy, H., Bhattacharjee, D.: Face sketch-photo matching using the local gradient fuzzy pattern. IEEE J. Intell. Syst. **31**(3), 30–39 (2016)
29. Roy, H., Bhattacharjee, D.: A novel quaternary pattern of local maximum quotient for heterogeneous face recognition. Elsevier Pattern Recogn. Lett. (2017). https://doi.org/10.1016/j.patrec.2017.09.029
30. Roy, H., Bhattacharjee, D.: A novel local wavelet energy mesh pattern (LWEMeP) for heterogeneous face recognition. Elsevier Image Vis. Comput. **72**, 1–13 (2018). https://doi.org/10.1016/j.imavis.2018.01.004
31. Roy, H., Bhattacharjee, D.: A ZigZag pattern of local extremum logarithm difference for illumination-invariant and heterogeneous face recognition. Springer Trans. Comput. Sci. **XXXI**, 1–19 (2018). https://doi.org/10.1007/978-3-662-56499-8_1
32. Peng, C., Gao, X., Wang, N., Li, J.: Graphical representation for heterogeneous face recognition. IEEE Trans. Pattern Anal. Mach. Intell. **xxx**, 1–13 (2016)
33. Sinha, P., Balas, B., Ostrovsky, Y., Russell, R.: Face recognition by humans: nineteen results all computer vision researchers should know about. In: Proceedings of IEEE, vol. 94 (2006)
34. Roy, H., Bhattacharjee, D.: Face sketch-photo recognition using local gradient checksum: LGCS. Springer Int. J. Mach. Learn. Cybern. xx (x), 1–13 (2016)
35. Ojala, T., Pietikinen, M., Menp, T.: Multiresolution gray-scale and rotation invariant texture classification with local binary patterns. IEEE Trans. Pattern Anal. Mach. Intell. **24**(7), 971–987 (2002)
36. Land, E.H., McCann, J.J.: Lightness and Retinex theory. J. Opt. Soc. Am. **61**(1), 1–11 (1971)
37. Horn, B.K.P.: Robot Vision. MIT Press, Cambridge, MA, USA (2011)
38. Zhang, T., Tang, Y.Y., Fang, B., Shang, Z., Liu, X.: Face recognition under varying illumination using gradientfaces. IEEE Trans. Image Process. **18**(11), 2599–2606 (2009)
39. Lai, Z., Dai, D., Ren, C., Huang, K.: Multiscale logarithm difference edgemaps for face recognition against varying lighting conditions. IEEE Trans. Image Process. **24**(6), 1735–1747 (2015)
40. Fan, C.N., Zhang, F.Y.: Homomorphic filtering based illumination normalization method for face recognition. Elsevier J. Pattern Recogn. Lett. **32**, 1468–1479 (2011)
41. An, G., Wu, J., Ruan, Q.: An illumination normalization model for face recognition under varied lighting conditions. Elsevier J. Pattern Recogn. Lett. **31**, 1056–1067 (2010)
42. Wang, H., Li, S., Wang, Y.: Generalized quotient image. In: Proceedings of IEEE International Conference on Computer Vision and Pattern Recognition, vol. 2, pp. 498–505 (2004)
43. Belhumeur, P., Georghiades, A., Kriegman, D.: From few to many: Illumination cone models for face recognition under variable lighting and pose. IEEE Trans. Pattern Anal. Mach. Learn. **23**(6), 643–660 (2001)
44. Lee, K.C., Ho, J., Kriegman, D.: Acquiring linear subspaces for face recognition under variable lighting. IEEE Trans. Pattern Anal. Mach. Learn. **27**(5), 684–698 (2005)
45. Li, S., Yi, D., Lei, Z., Liao, S.: The CASIA NIR-VIS 2.0 face database. In: Proceedings IEEE International Workshop on Computer Vision and Pattern Recognition, pp. 348–353 (2013)
46. Liu, X., Song, L., Wu, X., Tan, T.: Transferring deep representation for NIR-VIS heterogeneous face recognition. In Proceedings of IEEE International Conference on Biometrics (2016)

3D Face Recognition Based on Volumetric Representation of Range Image

Koushik Dutta, Debotosh Bhattacharjee, Mita Nasipuri and Anik Poddar

Abstract In this paper, a 3D face recognition system has been developed based on the volumetric representation of 3D range image. The main approach to build this system is to calculate volume on some distinct region of 3D range face data. The system has mainly three steps. In the very first step, seven significant facial landmarks are identified on the face. Secondly, six distinct triangular regions A to F are created on the face using any three individual landmarks where nose tip is common to all regions. Further 3D volumes of all the respective triangular regions have been calculated based on plane fitting on the input range images. Finally, KNN and SVM classifiers are considered for classification. Initially, the classification and recognition are carried out on the different volumetric region, and a further combination of all the regions is considered. The proposed approach is tested on three useful challenging databases, namely Frav3D, Bosphorous, and GavabDB.

Keywords Volumetric representation · 3D range image · Facial landmark
Plane fitting · Classification

K. Dutta (✉) · D. Bhattacharjee · M. Nasipuri
Department of Computer Science and Engineering, Jadavpur University, Kolkata, West Bengal, India
e-mail: koushik.it.22@gmail.com

D. Bhattacharjee
e-mail: debotoshb@hotmail.com

M. Nasipuri
e-mail: mitanasipuri@gmail.com

A. Poddar
Computer Science Engineering, Birla Institute of Technology, Mesra, Ranchi, India
e-mail: apoddar2008@gmail.com

© Springer Nature Singapore Pte Ltd. 2019
R. Chaki et al. (eds.), *Advanced Computing and Systems for Security*,
Advances in Intelligent Systems and Computing 883,
https://doi.org/10.1007/978-981-13-3702-4_11

1 Introduction

In the recent world, security is one of the most important issues in various fields due to a large amount of data hacking. Nowadays various social networking sites like Facebook and Twitter interact with us deeply. From the bad side of this interaction various risks like data leaks, Trojan, and impersonation may occur, where the personal information can be easily hacked by password hacking. On the other side, sometimes password and pin number are tough to properly memorize by us. To solve these types of issue, biometric-based verification and recognition system is used, which uses physical characteristic-based data. There are various biometrics like iris, fingerprints, face, and DNA. Among all of these verification or recognition system, face recognition system has taken significant attention from the researchers for the last few decades due to its various application domain like security, access control, and law enforcement. Face recognition [1] is possible in both 2D and 3D domain. 3D face recognition eliminates the problem like illumination and poses variation. As well as, working with 3D data gives extra geometrical information of face than a 2D face image. The methods are used for face recognition like holistic-based, feature-based, and hybrid-based approaches. The well-known holistic-based approaches are PCA [2, 3], LDA [4], and ICA [5]. The feature-based approaches include the extracted features from curvature analysis [6, 7] and structural matching-based features [8]. Finally some hybrid approaches [9, 10], which are basically the combination of holistic and feature-based approaches.

Feature extraction from images is very much challenging task in case of face recognition. It is difficult to extract innovative features from range images that recognize or verify person in the 3D domain. A large number of features can easily detect any person. Our main objective of this work is to represent the range image in a different form and also minimize the number of features that represent the face reliably. The alternative representation is essentially capturing the surface information.

The present system of 3D face recognition introduces a new approach for generating features by calculating the volume of the selected triangular regions on input range face images. All the significant contributions of this proposed work are enlisted below:

- Seven distinct landmarks are identified on significant portions like nose, eye, and mouth region of the input range face. Initially, the pronasal is chosen at the position, where curvedness and mean curvature both are maximum of multiple highest depth position. Next using the pronasal, other landmarks like eye and mouth corners are detected using the geometrical calculation of the face followed by mask representation for better localization.
- From those seven detected landmarks, three distinct landmarks are considered for the creation of one triangular region. In this way, six separate triangular regions are established.

- Next, the volume is calculated of all the triangular regions that produce a volumetric representation of the 2.5D face image. The volume is calculated by fitting plane on 3D range face image or 2.5D image.
- Finally, recognition accuracy of individual regions followed by merging of all separated regions is calculated successfully.

Volumetric representation of 3D range image is a new approach for calculating recognition rate. There are various other similar works that transform the original input range image to another form of 3D voxel representation and also other different formation. The voxel representation is basically the transformation of the original 3D mesh into a discrete and regular representation of a fixed number of volume elements or voxels. In [11], authors are presented new 3D voxel-based representation for face recognition. Here they have extracted three kinds of voxel representations: (1) those consisting of a single cut of the cube; (2) those consisting of a combination of single cuts of the cube; (3) depth maps. Support vector machine and PCA with Euclidean distance are used for matching the testing result. Pose and expression variant input images are considered for this work. Next in [12], the authors present a descriptor named as speed up local descriptor (SULD) from significant points extracted from the range image. They use frontal as well as pose variation images for the inputs of the proposed model. Significant points have been extracted by using Hessian-based detector. The SULD descriptor against all the significant points is computed for features creation. Another work in [13], where the authors have proposed a system where three different regions: eye, nose, and mouth separately classify the image. Initially, the whole range face image transforms to its local binary pattern (LBP) form. Further histogram of oriented gradient (HOG) is used for feature extraction. Here neutral, occlusion, and expression invariant images are considered as input.

The paper is organized as follows: Sect. 2 illustrates different steps of the proposed system. Next, Sect. 3 presents the result of our proposed work. After that Sect. 4 illustrates the comparative analysis with previous techniques. Conclusion and future work are given in Sect. 5.

2   The Present Proposed System

The present system is divided into five different stages. The stages are listed below:

- 3D Range Image Acquisition
- Smoothing
- Landmark Identification
- Volumetric Representation
- Classification

2.1   3D Range Image Acquisition

The range image is also named as a 2.5D image [14] or depth image. It contains depth value in the normalized range in between 0 and 255. In the range image, Z or depth value is mapped in the X-Y plane. Considering depth value as intensity, it looks like the gray level image. The range image is constructed from the 3D point cloud, which is captured by the 3D scanner. During this investigation, authors are considered range images of Frav3D, Bosphorous, and Gavab databases.

2.2   Smoothing

After capturing 3D point clouds, it consists of various noises like spike and holes. For removing spike from the range image, various different 2D filters can be used like: Gaussian smoothing filter, mean filter, median filter, and weighted median filter. Here the weighted median filter [15] is used for removing spike. The weighted median filter is same as a generalized median filter, where the mask of filter consists of a nonnegative integer weighted value.

2.3   Landmark Detection

In our work, landmark identification is a very crucial step for the creation of triangular region over the face. A landmark is a point which has a biological meaning that can identify faces differently. Generally, two types of a landmark can be detected. Hard-tissue landmarks lie on the skeletal and soft-tissue landmarks are on the skin and can be identified on the 3D point clouds or on range images. In human face total, 59 landmarks could be collected, but 20 of them are most famous.

Here, all total seven significant landmarks are considered such as pronasal (nose tip), left and right endocanthions (inner eye corner), left and right exocanthions (outer eye corner), and left and right chelions (mouth corner). All the detected landmarks cover important regions of the face. Various techniques are already implemented for landmark detection [16]. Here, for localization of the landmarks, different shape descriptors like coefficients of fundamental forms, mean and Gaussian curvature, maximum and minimum curvature, shape index and curvedness index are used. Here, for the nose tip detection, maximum depth measurement followed by curvedness index, mean curvature is used. The mean (K) and Gaussian curvature (H) is calculated from the first and second order derivative of input depth matrix as shown in Eq. 1. Further, curvedness index (CI) is calculated from the maximum (k_1) and minimum (k_2) curvatures shown in Eq. 2. P is a differentiable function $z = P(x, y)$, P_x is the first derivative of P with respect to x, and P_y is the first derivative of P with respect

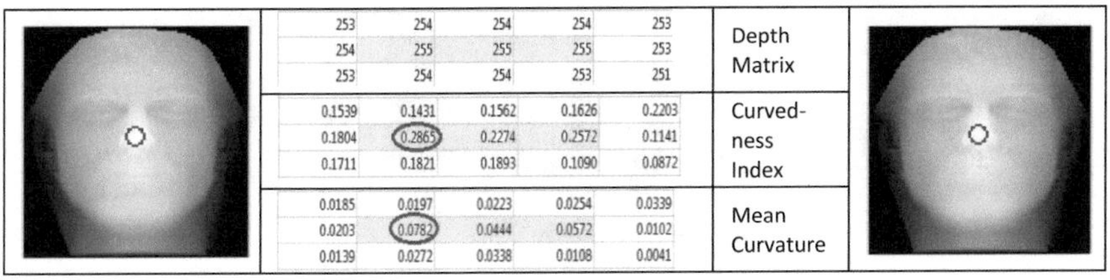

253	254	254	254	253	Depth Matrix
254	255	255	255	253	
253	254	254	253	251	
0.1539	0.1431	0.1562	0.1626	0.2203	Curved-ness Index
0.1804	0.2865	0.2274	0.2572	0.1141	
0.1711	0.1821	0.1893	0.1090	0.0872	
0.0185	0.0197	0.0223	0.0254	0.0339	Mean Curvature
0.0203	0.0782	0.0444	0.0572	0.0102	
0.0139	0.0272	0.0338	0.0108	0.0041	

Fig. 1 Nose tip localization

to y. P_{xx} is the second derivative of P with respect to x, P_{yy} is the second derivative of P with respect to y and, finally, P_{xy} is the mixed derivative.

$$K = \frac{P_{xx}P_{yy} - P_{xy}^2}{\left(1 + P_x^2 + P_y^2\right)^2}; \quad H = \frac{\left(1 + P_x^2\right)P_{yy} - 2P_x P_y P_{xy} + \left(1 + P_y^2\right)P_{xx}}{\left(1 + P_x^2 + P_y^2\right)^{3/2}} \quad (1)$$

$$CI = \sqrt{\frac{k_1^2 + k_2^2}{2}} \quad \text{where} \quad k_1 = H + \sqrt{H^2 - K}; k_2 = H - \sqrt{H^2 - K} \quad (2)$$

Next, the geometric structure of the human face is considered for eye and mouth corner detection according to [17]. After the nose tip identification, rest of the landmarks are identified on the basis of nose tip. There is a geometrical relationship between nose tip and eye corners as well as mouth corners. Here, a circle is considered of radius say R around the nose tip point (a, b) as a center. The geometrical formulae for the detection of all other landmarks against the nose tip are shown in Eq. 3. Here, Ø is the angle between nose tip and other landmark point and (x, y) is the point of the proposed landmark.

$$x = R\cos\theta + a; \quad y = R\sin\theta + b \quad \text{where} \quad \theta = (pi/180) * \text{Ø} \quad (3)$$

After finding the points, in some of the cases the detected points are not accurate. So, the authors are concentrated for accurate localization for the landmarks. A mask of 7 × 7 is considered around the point and then examined all types of surface descriptor and derivative values on that mask. It is finding a way for appropriate localization of mouth and eye corners. The details of the landmark detection are given below.

Pronasal: Nose tip or pronasal is one of the most important facial landmark points, where depth value is maximum. In some cases, maximum depth value occurs multiple times in connected pixels of depth matrix. In that case, a new approach has taken to calculate curvedness index and mean curvature on those pixels and consider the point where both curvedness index and mean curvature are maximum. In Fig. 1, red mark identifies the point, where curvedness index and mean curvature are maximum.

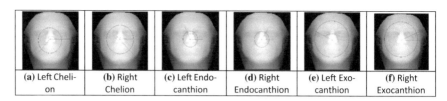

| (a) Left Cheli-on | (b) Right Chelion | (c) Left Endo-canthion | (d) Right Endocanthion | (e) Left Exo-canthion | (f) Right Exocanthion |

Fig. 2 Mouth, inner and outer eye corner localization

Chelion: Mouth corners or chelion are the landmark locations on the mouth. According to [17], consider a circle of radius r ($15 \leq r \leq 25$) pixels drawn around the center on nose tip. Further, left mouth corner located at the angle θ_1 ($220 \leq \theta_1 \leq 240$) left of the nose tip on the circumference and right mouth corner located at the angle θ_2 ($300 \leq \theta_2 \leq 320$) on the circumference. Figure 2a, b shows the left and right mouth corner landmark.

After getting the left and right mouth corner points, optimize the point consider 7×7 mask around the point.

Step 1: Consider the points, whose shape index belongs to range of Saddle surface, i.e., $[-0.125, 0.125)$

Step 2: Take the points where the first order derivative of z with respect to x greater than zero for left mouth corner and less than zero for right mouth corner

Step 3: The minimum of g of first fundamental form gives the left and right mouth corners

---

Endocanthion: Inner eye corners or endocanthions are the points where upper and lower inner eyelid meets. According to [17], a circle of radius r ($10 \leq r \leq 15$) draws around the center on nose tip. Further, left mouth corner located the angle θ_1 ($110 \leq \theta_1 \leq 130$) left of the nose tip on the circumference and right mouth corner located at the angle θ_2 ($50 \leq \theta_2 \leq 70$) on the circumference. Figure 2c, d shows the left and right inner eye corner landmark.

Similarly, after getting the left and right inner eye corner points, the following steps are considered for optimization of point using 7×7 mask around the point.

Step 1: Consider the points, whose shape index belongs to range of Ridge surface, i.e., $[0.375, 0.625)$

Step 2: Take the points where the first order derivative of z with respect to x greater than zero for both right and left eye

Step 3: The coefficients f and F of first and second fundamental forms greater than zero for the left eye and less than zero for the right eye

Step 4: The minimum of e of first fundamental form gives the left and right eye corner points

---

Fig. 3 Triangular regions
against detected landmarks

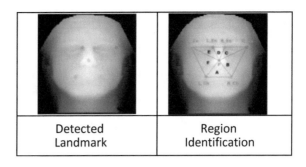

Detected Landmark	Region Identification

Exocanthion: Outer eye corners or exocanthions are the points where upper and lower outer eyelid meets. According to [17], consider a circle of radius r $(25 \leq r \leq 35)$ pixels drawn around the center on nose tip. Further, left outer eye corner is located at approximately $\theta_1 (150 \leq \theta_1 \leq 170)$ left of the nose tip on the circumference and right outer eye corner is located the angle $\theta_2 (10 \leq \theta_2 \leq 30)$ on the circumference. Figure 2e, f shows the left and right outer eye corner landmark.

Similar as before, after getting the left and right outer eye corner points, the following steps are considered for optimization of point using 7×7 mask around the point.

Step 1: Consider the points, whose shape index belongs to the range of Rut surface, i.e., $[-0.625, -0.375)$

Step 2: Take the points where the first order derivative of z with respect to x greater than zero for both right and left eye

Step 3: Divide the points of the first order derivative of z with respect to y into two clusters using K-means clustering technique. Consider the cluster of lower center value

Step 4: The points where Gaussian curvature K equal to zero and mean curvature less than zero

Step 5: The minimum of e of first fundamental form gives the left and right eye corner points

2.4 Volumetric Representation

This section is divided into two subsections: Triangular region identification and volume calculation.

Triangular Region Identification. After the landmark detection stage, the triangular region has been created based on three individual landmarks. According to the seven significant landmarks, all total six triangular regions: A, B, C, D, E, and F have built. The triangular region can be identified as from Fig. 3:

- Region A: ∇PL_ChR_Ch
- Region B: ∇PR_ChR_Ex

Fig. 4 Volume calculation
on separate regions

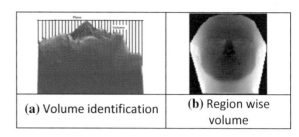

| **(a)** Volume identification | **(b)** Region wise volume |

- Region C: $\nabla P R\_En R\_Ex$
- Region D: $\nabla P L\_En R\_En$
- Region E: $\nabla P L\_Ex L\_En$
- Region F: $\nabla P L\_Ch L\_Ex$

Here 'P' denotes pronasal, L_Ch and R_Ch denote left and right chelion, L_En and R_En denote left and right endocanthions, and L_Ex and R_Ex denote left and right exocanthions. The DDA line drawing [18] algorithm is used for the creation of line between two distinct landmark points on range face image. All total six triangular regions cover a maximum portion of the face. Figure 3 illustrates all the triangular regions against different landmarks.

Volume Calculation. Volumetric representation gives the volume of the portion of the range face image. Volume is calculated on the triangular region A, B, C, D, E, and F separately. To measure the volume, at first a plane is fitted at the position of maximum depth (i.e., nose tip) of the range image. After plane fitting, the authors have calculated the distance between the plane and all other pixels of the face region as shown in Fig. 4a. The distance between maximum depth plane and other pixels is also treated as height or density of that pixel. Further, the summation of density is calculated the volume of any particular region. The resultant inverse range image is also named as density range image that gives the volume of the image. The volume of the entire separate region shown in Fig. 4b is calculated in the same way. Volume is calculated on all the six distinct regions, which is identified as features of any subject face.

Volume calculation of triangular region of density range image is done based on the scanline approach of computer graphics. Moreover, considering the triangular region, three information needed to be known for volume calculation of any region:

- Choosing of starting point
- Choose row-wise or column-wise scanline approach
- Farthest point detection in between other two points.

In our work, the authors are considered nose tip point or pronasal point of the face as the starting point because this point is common in all the triangular regions. Three cases according to the x and y position of the points occur before applying scanline approach as shown in Fig. 5. From the figure, according to the image axis if the x value of two other points (excluding starting point) is on the same side, it

Fig. 5 Different cases of scanline technique

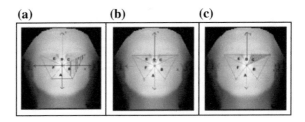

(a)   (b)   (c)

may be increasing or decreasing, and y value is on different sides than considering column-wise scanline approach as shown in Fig. 5a. Similarly, when the y value of two other points (excluding starting point) is on the same side, it may be increasing or decreasing, and x value is on different sides than considering row-wise scanline approach as in Fig. 5b. In some cases, when both x and y values are either increasing or decreasing than any approach can be considered as in Fig. 5c.

Another problem may also arise at the time of calculation that is the number of the index of the line array is not same as x value or y value difference. Due to the aliasing effect of a line drawing of the triangular region this problem has occurred. Now, the steps for volume calculation of any triangular region are given below.

Step 1: Take three points of the triangular region of the density range image
Step 2: Spatial coordinate vectors of line drawing between two points are taken
Step 3: Choose the starting point according to the problem. Here, pronasal is considered
Step 4: Check the position of the other two points. After that choose the scanline approach: row wise or column wise
Step 5: Add the density value according to one scanline. Then jump to next scanline and then one after another. In this way add the total density values. After this addition, the volume of any triangular region is calculated

---

The proposed work of volumetric representation can be mathematically justified by volume integral. It can be usually denoted as in Eq. 4.

$$\int_V f(x, y, z)d\tau \tag{4}$$

Here $f(x, y, z)$ denoted as z value of (x, y) location of the negative range image, which is basically identified as distance between highest depth plane and any particular pixel. Function is basically the summation of all pixels in a particular region. Let a, b, and c are the sides of the triangle and A denotes area of the circle. Equations 5 and 6 illustrate volume of the triangular region in this work.

$$A = \sqrt{S(S-a)(S-b)(S-c)} \quad \text{where } S = \frac{(a+b+c)}{2} \tag{5}$$

Table 1 Details of input databases

Database	Year	No. of subjects	No. of scans	Variation
Bosphorous [19]	2008	105 (60-M, 45-F)	4652	Pose, expression, occlusion
FRAV3D [20]	2006	106	1696	Pose, expression, illumination
GavabDB [21]	2004	61 (45-M, 16-F)	427	Smile, frontal accentuated laugh, frontal random gesture

$$V = \int_{Z_1}^{Z_2} A dz, \text{ where } Z_1 \text{ and } Z_2 \text{ are density value} \qquad (6)$$

2.5 Classification

For classification, support vector machine (SVM) and K-nearest neighbour (KNN) classifiers were used. For the KNN, Euclidean distance is used for nearest distance calculation. Here twofold cross-validation technique is applied on input dataset, where the whole dataset is divided into two separate regions: training and test set.

3 Experiment and Result

The proposed work is tested on range images of three well-known databases Frav3D, Bosphorous, and GavabDB. The details of these databases are given below in Table 1.

The 3D data of these databases have been captured by the available 3D scanner. Initially 3D point cloud, further it is transformed to range images. Here in our work, only frontal faces including neutral and expression variation range images of size 100×100 are considered. At first, the accuracy of landmark localization compared to Bagchi et al. [17] is illustrated in Table 2. The standard deviation error (SDE) of seven distinct landmark localization of our proposed method and Bagchi et al. work comparable with the ground truth is represented. The 4 neutral and 2 expression variation images of FRAV3D, 2 neutral, 6 expression, 20 lower face AU (Action Units) (LFAU), 5 upper face AU (UFAU) and 2 Combined AU (CAU) of Bosphorous and 2 neutral and 2 expression face images of Gavab database are considered for landmark localization and further recognition. Here the recognition accuracy is calculated on separate regions. Further, the average of all distinct region's recognition accuracy correspond to the recognition accuracy of whole face region. Tables 3 and 4 given below illustrate the recognition rates of three databases using KNN and SVM classifier.

Table 2 Standard Deviation Error of proposed landmarks

Database	Variation	SDE (Proposed work)							SDE (Bagchi et al. [17])						
		P	L_Ch	R_Ch	L_En	R_En	L_Ex	R_Ex	P	L_Ch	R_Ch	L_En	R_En	L_Ex	R_Ex
FRAV3D	Neutral	0.33	4.24	2.5	3.14	3.78	2.27	3.98	0.35	3.98	2.13	4.36	3.53	3.43	5.66
	Expression	0.5	6.82	4.42	3.14	3.32	1.66	2.54	0.61	6.6	5	3.28	3.67	3.12	3.79
BOSPHOROUS	Neutral	0.6	3.65	4.72	4.77	4.23	2.3	3.1	0.65	3.2	5.21	5.32	4.78	4.56	4.13
	Expression	0.55	5.41	6	3.28	2.89	4.3	2.98	0.4	5.74	6	4.87	4	5.47	4.23
	CAU	0.32	6.5	5.55	4.1	3.7	2.16	3.67	0.3	6.8	5.28	5.3	4.23	2.35	4.79
	LFAU	0.53	6.3	4.56	3.12	2.59	3.3	2.98	0.6	7.1	4.23	4.22	4	3.9	3.14
	UFAU	0.2	5.33	3.88	2.73	3	5.77	3.21	0.22	5	4.1	3.64	3.11	7.31	4.68
GAVAB	Neutral	0.3	3.86	2.56	2.87	3.7	1.5	2.34	0.32	4	2.44	3.11	3.72	3.2	3.35
	Expression	0.41	4.25	3	5.48	4.71	2.11	3.19	0.38	4.75	4.1	7.7	5.23	4.15	4.63

Table 3 Recognition accuracy of three input databases using KNN classifier

Database	Region A	Region B	Region C	Region D	Region E	Region F	Average
Frav3D	94.4	93.8	94	94.9	94.6	94	94.28
Bosphorous	95.88	95	94.77	95.45	95	95.2	95.21
GavabDB	91.68	90.3	91	90.6	91.4	90	90.83

Table 4 Recognition accuracy of three input databases using SVM classifier

Database	Region A	Region B	Region C	Region D	Region E	Region F	Average
Frav3D	95.58	95.3	96	95.9	95.78	95	95.59
Bosphorous	96.87	96	96.7	96	96.5	96.2	96.37
GavabDB	93.68	91.3	92	93.8	91.89	92.4	92.51

Table 5 Comparison of recognition performance of the proposed method with some other methods on the Frav3D database

Methods	Accuracy (%)	References
Curvature analysis + SVD + ANN (Classification on the whole face)	86.51	Ganguly et al. [6]
LBP + HOG + KNN (Region-based classification)	88.86	Dutta et al. [13]
Geodesic texture wrapping + Euclidean-based classification (Classification on the whole face)	90.3	Hajati and Gao [22]
ICP-based registration + Surface Normal + KPCA	92.25	Bagchi et al. [10]
Proposed Method (Triangular representation + Volume calculation + KNN)	**94.28**	This paper
DWT + DCT + PCA + Euclidean distance classifier	94.50	Naveen and Mon [23]
Proposed Method (Triangular representation + Volume calculation + SVM)	**95.59**	This paper

4   Comparative Study and Analysis

In our proposed work, Frav3D, Bosphorous, and Gavab databases are used as input database. There are already few works have already done on these databases. The comparison studies with previous other tasks of the three different databases: Frav3D, Bosphorous, and Gavab have been discussed in Tables 5, 6, and 7, respectively.

The recognition accuracies of our proposed method on Frav3D, Bosphorous, and Gavab database are 94.28%, 95.21%, and 90.83%, respectively, using KNN classifier and 95.59%, 96.37%, and 92.51%, respectively, using SVM classifier. Compare with other works, it can be established that our proposed work has not covered whole face region, whereas the existing proposed works of different authors had worked with whole face region. On the other side, the size of the feature set is short compared to other methods. On that basis it can say, the proposed system is more accurate.

Table 6 Comparison of recognition performance of the proposed method with some other methods on the Bosphorous database

Methods	Accuracy (%)	References
ICP-based recognition	89.2	Dibeklioglu et al. [24]
Proposed Method (Triangular representation + Volume calculation + KNN)	**95.21**	This paper
ICP-based holistic approach + Maximum likelihood classifier	95.87	Alyuz et al. [25]
Surface Normal and KPCA	96.25	Bagchi et al. [10]
Proposed Method (Triangular representation + Volume calculation + SVM)	**96.37**	This paper

Table 7 Comparison of recognition performance of the proposed method with some other methods on the Gavab database

Methods	Accuracy (%)	References
Mean and Gaussian curvature-based Segmentation	77.9	Moreno et al. [26]
Geometrical feature + PCA versus SVM classification	90.16	Moreno et al. [3]
Proposed Method (Triangular representation + Volume calculation + KNN)	**90.83**	This paper
2DPCA + SVM classifier	91	Mousavi et al. [27]
Proposed Method (Triangular representation + Volume calculation + SVM)	**92.51**	This paper

5 Conclusion

Here a new technique has been presented for 3D face recognition, where a new representation of range images has produced by calculating the volume of the regions of the face. Initially, landmarks are identified in an optimum way. The further triangular region has constructed by calculating line DDA algorithm. Next, a new algorithm has presented for calculating the volume of the selected triangular regions. The volumes of all distinct regions are used to construct feature vectors for classification. Overall it can be concluded that the volume representation of range images is a new approach in the 3D domain. The system is computationally efficient for recognition with high recognition accuracy. Frontal faces including neutral and expression variant range images of three well-known databases are used as input to this system. In the future, pose variant images will be considered for this technique.

Acknowledgements I, Koushik Dutta, would like to express thanks to Ministry of Electronics and Information Technology (MeitY), Govt. of India. I am also thankful for Department of Computer Science and Engineering, Jadavpur University, Kolkata, India for providing the necessary infrastructure for this work.

References

1. Abate, F., Nappi, M., Riccio, D., Sabatino, G.: 2D and 3D face recognition: a survey. Pattern Recogn. Lett. **28**(14), 1885–1906 (2007)
2. Gervei, O., Ayatollahi, A., Gervei, N.: 3D face recognition using modified PCA methods. World Acad. Sci. Eng. Technol. **4**(39), 264 (2010)
3. Moreno, A.B., Sanchez, A., Velez, J.F., Diaz, J.: Face recognition using 3D local geometrical features: PCA vs SVM. In: Proceedings of the ISPA, pp. 185–190 (2005)
4. Heseltine, T., Pears, N., Austin, J.: Three-dimensional face recognition: a fisher surface approach. In: Proceedings of the ICIAR, pp. 684–691 (2008)
5. Hesher, C., Srivastava, A., Erlebacher, G.: A novel technique for face recognition using range imaging. In: Proceedings of the Seventh International Symposium on Signal Processing and Its Applications, pp. 201–204 (2003)
6. Ganguly, S., Bhattacharjee, D., Nasipuri, M.: 3D face recognition from range images based on curvature analysis. ICTACT J. Image Video Process. **4**(03), 748–753 (2014)
7. Ganguly, S., Bhattacharjee, D., Nasipuri, M.: Fuzzy matching of edge and curvature based features from range images for 3D face recognition. Intell. Autom. Soft Comput. (IASC) **23**(1), 51–62 (2017)
8. Berretti, S., Werghi, N., Bimbo, A., Pala, P.: Matching 3D face scans using interest points and local histogram descriptors. Special section on 3D object retrieval. Comput. Graph. **37**, 509–525 (2013)
9. Chouchane, A., Belahcene, M., Ouamane, A., Bourannanae, S.: 3D face recognition based on histogram of local descriptors. In: 4th International Conference on Image Processing Theory, Tools and Applications (IPTA), pp. 1–5 (2014)
10. Bagchi, P., Bhattacharjee, D., Nasipuri, M.: 3D face recognition using surface normals. In: TENCON 2015—2015 IEEE Region 10 Conference, pp. 1–6 (2015)
11. Moreno, A.B., Sanchez, A., Velez, J.F.: Voxel-based 3D face representations for recognition. In: 12th International Workshop on Systems, Signals and Image Processing, pp. 285–289 (2005)
12. Shekar, B.H., Harivinod, N., Kumara, M.S., Holla, K.R.: 3D face recognition using significant point based SULD descriptor. In: International Conference on Recent Trends in Information Technology (ICRTIT) (2011)
13. Dutta, K., Bhattacharjee, D., Nasipuri, M.: Expression and occlusion invariant 3D face recognition based on region classifier. In: 1st International Conference on Information Technology, Information Systems and Electrical Engineering (ICITISEE), pp. 99–104 (2016)
14. Ganguly, S., Bhattacharjee, D., Nasipuri, M.: 2.5D face images: acquisition, processing and application. In: ICC 2014—Computer Networks and Security, pp. 36–44 (2014)
15. Yin, L., Wang, R., Neuvo, Y.: Weighted median filters: a tutorial. IEEE Trans. Circuits Syst. 11: Analog Digit. Signal Process. **43**(3) (1996)
16. Ahdid, R., Taifi, K., Safi, S., Manaut, B.: A survey on Facial Features Points Detection Techniques and Approaches. Int. J. Comput. Electr. Autom. Control Inf. Eng. **10**(8), 1566–1573 (2016)
17. Bagchi, P., Bhattacharjee, D., Nasipuri, M.: A robust analysis, detection and recognition of facial features in 2.5D images. Multimed. Tools Appl. **75**, 11059–11096 (2016)
18. DDA Line Drawing. http://www.geeksforgeeks.org/dda-line-generation-algorithm-computer-graphics/
19. Bosphorous. http://bosphorus.ee.boun.edu.tr/default.aspx
20. FRAV3D. http://www.frav.es/databases
21. GavabDB. http://gavab.escet.urjc.es/recursos_en.html
22. Hajati, F., Gao, Y.: Pose-invariant 2.5D face recognition using geodesic texture warping. In: 11th International Conference on Control, Automation, Robotics and Vision Singapore, pp. 1837–1841 (2010)
23. Naveen, S., Moni, R. S.: A robust novel method for Face recognition from 2D depth images using DWT and DCT fusion. In: International Conference on Information and Communication Technologies (ICICT), pp. 1518–1528. Elsevier (2014)

24. Dibeklioglu, H., Gokberk B., Akarun, L.: Nasal region-based 3D face recognition under pose and expression variations. In: ICB'09: Proceedings of the Third International Conference on Advances in Biometrics, pp. 309–318. Springer, Berlin, Heidelberg (2009)
25. Alyuz, N., Gokberk, B., Akarun, L.: A 3D face recognition system for expression and occlusion invariance. In: BTAS'08: Proceedings of the IEEE Second International Conference on Biometrics Theory, Applications and Systems, Arlington, Virginia, USA (2008)
26. Moreno, A.B., Sanchez, A., Velez, J.F., Diaz, J.: Face recognition using 3D surface-extracted descriptors. In: Irish Machine Vision and Image Processing Conference (2003)
27. Mousavi, M.H., Faez, K., Asghari, A.: Three dimensional face recognition using SVM classifier. In: Seventh IEEE/ACIS International Conference on Computer and Information Science, Portland, pp. 208–213 (2008)

Author Index

© Springer Nature Singapore Pte Ltd. 2019
R. Chaki et al. (eds.), *Advanced Computing and Systems for Security*,
Advances in Intelligent Systems and Computing 883,
https://doi.org/10.1007/978-981-13-3702-4

Printed in the United States
By Bookmasters